The Health Commons:

A Handbook of Nursing Practice

Health Commons

Created by Kathleen M. Clark on behalf of the Augsburg College Nursing Department

INTENTION OF THE HANDBOOK

This handbook has been created for multiple reasons. Most importantly, we wanted to create a manual that would allow for students to gain further insights into their experiences at the Health Commons. This means that students can use reflections and activities in this publication to gain deeper understanding into the complexities involved with caring for those who are marginalized in our society. For example, a student has limited time to reflect one-on-one with a faculty member about all the events that occur in a Health Commons experience. This handbook would allow students time to reflect and apply the practice model and 'Rules of Thumb' to new experiences. Or, the handbook can serve as a way for students to prepare for an upcoming scheduled practicum.

The second reason for creating this handbook is for those nursing students, organizations or community members who are interested in creating a Health Commons in their local communities. This publication is based on the experiences and practice knowledge of faculty and students of the Augsburg College nursing programs. It is compiled as a guide for people to gain further knowledge about care being provided at the Health Commons and what is needed to adapt the model in new settings. Chapter 5 discusses starting a Health Commons in detail once you have worked your way through the readings and activities at the beginning of the manual to first begin understanding the practice model and 'Rules of Thumb'.

We are honored by the number of people who have inquired about learning more regarding the Health Commons. This handbook has been intentionally created using 'common' language in order to allow for anyone, not just a nurse, to begin to understand the complexities that are involved when working in a community. Our hope is that through the experiences and relationships at the Health Commons, people will be able to challenge stereotypes and biases that keep the community from growing stronger and create inequity.

Thank you for taking the time to read about the work at the Health Commons. Please feel free to email or contact us with any questions you may have in understanding the model of care or work being done at the Health Commons. We will end this introduction in the words of Margaret Mead, "Never doubt that a small group of thoughtful, committed citizens can change the world. Indeed, it is the only thing that ever has."

-Augsburg College Department of Nursing

ACKNOWLEGMENTS

There are many people who have helped in the creation of this handbook in various ways. At this time, the Augsburg College Nursing Department would like specifically acknowledge the 2012 graphic design students at the college and their professor, Chris Houltberg, for their contribution of the images throughout this manual. These student created a gallery featuring the Health Commons depicting the practice model and stories in various visual images as well as created the logo. Thank you for your hard work, enthusiasm and creativity to help create a vibrant manual for the Health Commons.

Legal Disclaimer and Copyright

Copyright Notices © 2014

All rights reserved. No parts of this book may be reproduced, stored in a retrieval system, or transmitted in any form or by any means, electronic, mechanical, photocopying, recording, scanning, or otherwise without the prior written permission of the creator of this docment.

Disclaimer

All the material contained in this book is provided for educational and informational purposes only. No responsibility can be taken for any results or outcomes resulting from the use of this material.

While every attempt has been made to provide information that is both accurate and effective, the author does not assume any responsibility for the accuracy or use/misuse of this information.

The Health Commons: A Handbook of Nursing Practice

Table of Contents

INTENTION OF THE HANDBOOK 2

ACKNOWLEGMENTS 4

Legal Disclaimer and Copyright 5
    Copyright Notices © 2014 5
    Disclaimer 5

CHAPTER ONE: What is a Health Commons? 8
    Basic Overview of the Space 11
    The Model of Care 12

CHAPTER TWO: The History 17
    Augsburg Central Health Commons 17
    Health Commons in Cedar-Riverside Neighborhood 19

CHAPTER 3: Rules of Thumb 24
    #1 Rule of Thumb 25
        Activity 1. 26
        Challenge 1. 27
    #2 Rule of Thumb 29
        Case Study 1. 30
    # 3 Rule of Thumb 31
        Activity 2. 32
    #4 Rule of Thumb 33
        Challenge 2. 34
    #5 Rule of Thumb 36
        Activity 3. 36
    #6 Rule of Thumb 37
    #7 Rule of Thumb 38
    #8 Rule of Thumb 39

CHAPTER FOUR: The Stories 40
    Story 1. 40
        Application 1. 43
    Story 2. 45
        Application 2. 47
    Story 3. 48
        Application 3. 49

CHAPTER 5: Where to Begin 50
Worksheet 1. 53
Worksheet 2. 54
Worksheet 3. 56
Worksheet 4. 57
Worksheet 5. 58
Worksheet 6. 59

CHAPTER SIX: Words to Live By 60

CHAPTER SEVEN: References 61

CHAPTER EIGHT: Further Information 62

CHAPTER ONE: What is a Health Commons?

This manual describes nursing in two separate community contexts within a shared inner-city environment. Presenting the unity of practice in very diverse settings emphasizes the importance of local context and the benefit of de-emphasizing a professional-expert model. The Health Commons is a unique place because it is created in response to expressed felt need by community members in local settings. In simple terms, the Health Commons is a drop-in center that is focused on health and healing. Actions and decisions focus upon the principle of hospitality as visitors are greeted with warmth and openness. The space represents safety, relationships and mutual benefit. We believe all people at the Health Commons can participate in helping one another establish a desired state of health. There are no experts or hierarchies in place, but a community of people who care.

Establishing a Health Commons focuses on a local context that can be beneficial to the community in many ways. It is a place that is owned by those who utilize it. For example, if young mothers in the area are in need of diapers, Health Commons nurses will explore the resources available in order to either provide direct assistance or guide people to the appropriate resources. In this sense, the Health Commons can provide needed immediate support to participants.

Faculty, student nurses, staff and volunteers help facilitate the day-to-day activities at the Health Commons. This allows nurses and volunteers to gain understanding into the complexities of the lives of people in our world. For example, a nursing student from rural Minnesota may have the opportunity to interact and listen without interruptions to the story of a Somali immigrant. In these moments, a human connection is made and stereotypes or preconceived ideas are challenged.

The focus of the care at the Health Commons is to build meaningful relationships. Many times in the U.S. health care system, health care providers are unable to take the time to truly listen to the patient. Nurses continue to have additional tasks and procedures that are required in a day's work. As more technological advances monopolize the nurse's time, the patient's emotional wellbeing is often ignored. At the Health Commons, nursing students, faculty, staff and volunteers are able to learn first-hand the value of listening and the importance of building relationships, which will in turn impact their current nursing practice.

Many people are marginalized in our current healthcare system. Many people from various backgrounds and socioeconomic statuses feel stereotyped by health care providers and experience confusion over the complexities of the systems in place. Participants at the Health Commons shared their stories of frustrations when visiting with the doctor or nurse. For example, one gentleman said, "Every time I go to the ER they just act like I want something I don't need. I've seen nurses roll their eyes and look at one another when I come in the doors. I can't help it that I get sick a lot and have nowhere else to go." Another woman said, "The way they talk to you when you go to the doctor is like they think you are stupid. They talk to me like they are my parents. I have parents, thank you very much!"

The Health Commons environment allows persons with such experiences to discuss their frustrations and to re-establish trust with health care providers. The nurses and volunteers at the Health Commons listen with intention and provide emotional support. Nurses and students also identify the strengths of the person they are meeting with throughout conversation. For example, if a student is having a conversation with a man who is currently staying at an emergency shelter while continuing to find a job, the student will identify the visitor's strength in that situation. The student may say, "Wow, it amazes me that you are able to wake up everyday and continue to look for a job when you don't know where you will sleep each night, where you will shower, or how you will get to the job interview. That is remarkable."

The Health Commons: A Handbook of Nursing Practice

I'm So Glad Our Friends Are Here Today

*I'm truly blessed,
We have more that brings us together
than what separates us.*

*Today I expect to receive answers to prayers.
Ask and you shall receive.
Special treatment to everyone.
Treat them like a queen.*

*Encourage.
Stand behind your friends.
No matter what happens,
you give your friend your blessings.*

-Poem from the Women's Group

Basic Overview of the Space

The room that the Augsburg Central Health Commons occupies is a former first-grade schoolroom at Central Lutheran Church. It has large windows, an open space, and an informal décor. People typically enter the room from a hallway, come through for some basic supplies, and stay to speak with the nurse about any issues or concerns in comfortable chairs. Individuals often stay and enjoy a cup of coffee in the Health Commons, as it is a place of community gathering and friendship building.

The room itself is located in the basement of this large, urban church called the Restoration Center. The Restoration Center is an outreach center that provides basic services to those experiencing homelessness or poverty in Minneapolis. The area that houses the Restoration Center used to be a school and was reorganized to function as a resource center. During the same hours that the Augsburg Central Health Commons (ACHC) is open, there is also a free clothing store, job coaching, Internet use, one-on-one resource assistance, and a community meal.

As you can see in the pictures below, the room is a welcoming space with natural lighting, comfortable seating, and tea or coffee is readily available.

The Health Commons: A Handbook of Nursing Practice

    This picture is of a Health Commons located in the largest East African immigrant community of Minneapolis. Again, the space is open and in a familiar setting to the community. Easy access and familiarity are key aspects of deciding on a physical place to host a Health Commons. A group of women from the neighborhood decorated the space itself, which adds to the overall warmth and hospitality it provides. Again, as one can see from the photos, the Health Commons is not a medical center or a clinic. It is a safe place located in familiar context to those served. This allows people to feel more at ease, comfortable and secure.

The Health Commons: A Handbook of Nursing Practice

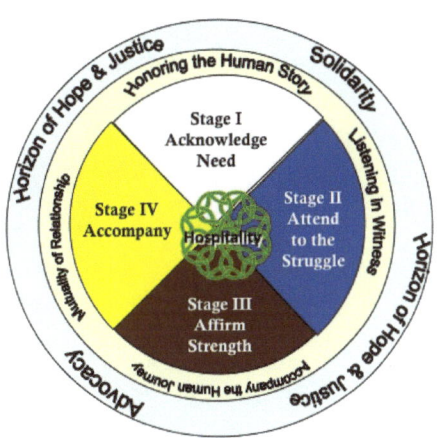

The Model of Care

The Health Commons model was developed in 2009 by a group of Doctor of Nursing Practice (DNP) students and faculty (Enestvedt, McHale, Miller, Loushin, Gunderson, Kinney, Freborg, Schuhmacher, & Baumgartner, 2009). The model was inspired by the experiences at the Health Commons in the downtown Minneapolis location, called the Augsburg Central Health Commons (ACHC), but can be applied to care provided in any location.

At the center of care at the Health Commons is hospitality. This focus is crucial when working with people who are marginalized and lack trust with health care providers. All those who enter the Health Commons are greeted warmly. For

example, in the Cedar-Riverside location, the Somali phrase "Soo dhowow", meaning "Welcome", is used to greet guests. Also, students, faculty, staff and volunteers are encouraged to dress culturally appropriately for the setting of the Health Commons for that particular day.

Hospitality also means the nurse must listen intently and be aware of any nonverbal communication cues that might be communicated. Eye contact and a smile are important tools to allow people to feel recognized and acknowledged. The room is set up in an intentional way and coffee or tea is readily available.

The first phase of the model is called 'acknowledge need' (Enestvedt et al., 2009). This process occurs when a participant at ACHC expresses a felt need for an item, such as socks or soap, which is provided to him or her without question. Also, acknowledgment is demonstrated at HCCR, when need for a community gathering space is identified and created. During this phase, building trust is the emphasis. This phase communicates care and compassion to a person in an unconditional way.

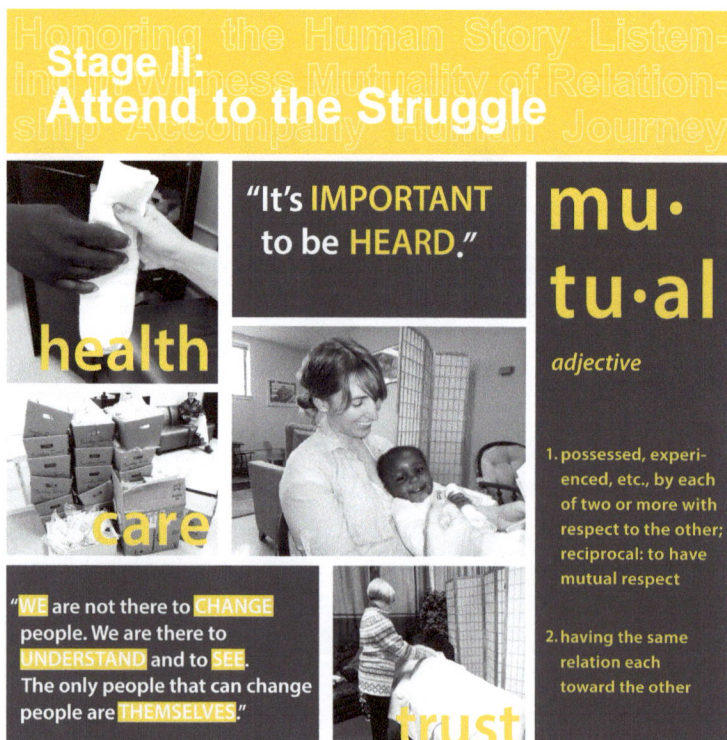

'Attend to the struggle' is the second phase of the Health Commons model (Enestvedt et al., 2009). This stage can be seen in many ways. At the ACHC, it typically occurs when the visitor asks the nurse for a simple action such as providing a blood pressure check or a Band-Aid. Usually participants in this phase of trust building are hesitant to share their personal struggles or life circumstances. This time of interaction represents the embodiment of the struggle of those who are marginalized in our world (Enestvedt et al., 2009).

This phase looks slightly different with participants at the HCCR. It usually is identified at a time when the conversation goes beyond the blood pressure check or other initial health related question that brought that person to the Health Commons. It's when someone briefly discusses his or her frustrations with utilizing the health care system or identifies a class that needs to be started such as knitting or yoga. The creating of such groups or classes is an additional way to attend to the struggle.

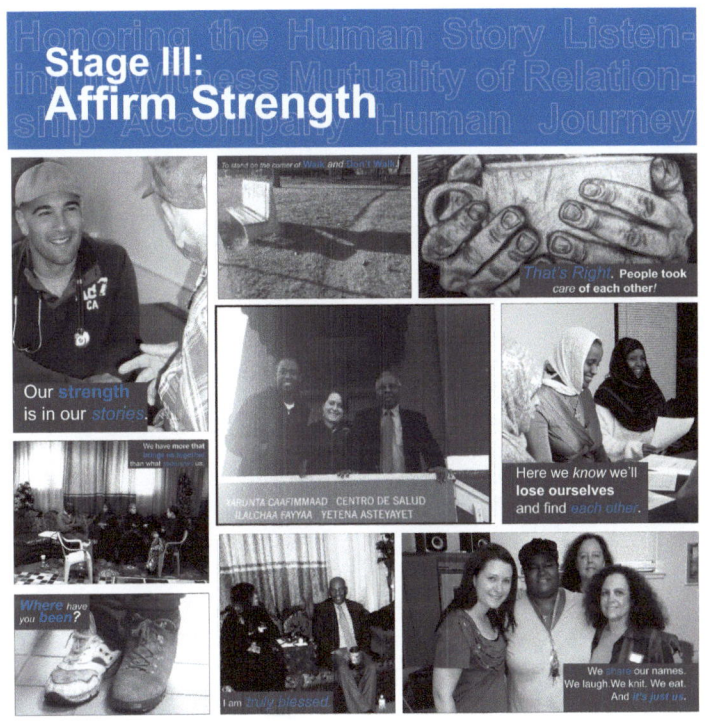

The third phase of the Health Commons model is 'affirm strength' (Enestvedt el al., 2009). This represents a place in time when a deeper relationship begins to develop with the nurse and participant. Story sharing begins and trust deepens to new levels of personal discourse. The nurse affirms the strengths of this person's story and acknowledges the resiliency demonstrated in his or her actions. The nurse must remember in this phase, and throughout the entire process, to shift from the expert model of Westernized health care, and to come together with everyone at the Health Commons as equals. At this time in the relationship building it is crucial that the nurse remembers to avoid trying to "fix" the situation for the visitor, but must focus on collaboration and mutuality (Enestvedt et al., 2009).

This is also a time to blend culturally appropriate needs, such as offering Black Seed, a Somali indigenous herbal practice, or allowing time for prayer. It is a time of learning what someone finds important to his or her health, and incorporating that into the interaction. People know their bodies and what helps them feel healthy: affirm that strength and answer questions along the way.

The final phase is called 'accompany.' (Enestvedt et al., 2009). This phase is demonstrated when a person who visits the Health Commons is able to obtain desired health goals through the relationships built with the nurses and students. To get to this stage, the participant will feel as though he or she is part of a community, has a place where he or she belongs, and is able to find the support he or she needs to create wanted outcomes (Enestvedt et al., 2009). In this phase, the nurse and the participant come together on their journey of partnership as they share friendship, support and the importance of being truly present with one another.

CHAPTER TWO: The History

HELEN WOELFEL, RN, coordinator of nursing center at Central, takes client's blood pressure.

Augsburg Central Health Commons

The first nursing-led drop-in center created by Augsburg College nursing faculty was called the Augsburg Central Nursing Center. Located at Central Lutheran Church in downtown Minneapolis, the center was developed over the course of many years. The chair of the Augsburg College Nursing Department at that time, Bev Nilssen, was a colleague of an associate pastor at Central Lutheran Church. Central Lutheran Church has a long history of being actively involved in serving the homeless population of the downtown area through service, outreach and community engagement. Dr. Nilssen wanted to develop an experience for nursing students to engage in a service-learning project by providing blood pressure checks and emotional support to those living on the streets. She also hoped to develop a practice site for nursing faculty. In October of 1992, the Augsburg College nursing department and staff at Central Lutheran Church partnered to open the Augsburg Central Nursing Center (currently called Augsburg Central Health Commons).

As the Nursing Center unfolded over time, the nursing practice focus evolved from health promotion in the community to health as community (Enestved et al., 2009). It became apparent that nursing students and faculty were not simply providing services to those living on the streets, but this was a place of mutual benefit and relationships. Students were able to learn of a culture that is hidden and often misunderstood in mainstream society. The sharing of life circumstances and survival skills of those visiting the center allowed students and faculty to gain insight into the life of those living in poverty. Through the process of relationship building, students and faculty were able understand the value of trust and openness when providing nursing care.

The ACHC has been open for over twenty years. In a typical week, over 120 people visit the ACHC. There are two streams of visitors who come to ACHC. The larger group comes for basic health protection items: socks, diapers, soaps, toothbrushes, toothpastes, razors, deodorants, lotions, shampoos, conditioners, combs, floss, adult incontinence supplies, feminine napkins, and condoms. Often, faces in the hygiene line become familiar and from time to time a so-far-nameless individual will take the next step in personal risk and ask to speak with a nurse. This is our second stream of visitors, which typically represents approximately thirty percent of those who visit us in a given day.

The ACHC has become a place that is part of the community. Our focus is to move towards principles of social justice instead of simple acts of charity. For example, we must hold people accountable in this space because this space belongs to everyone. Some of our participants have helped volunteer or have spoken to a group of students about his or her life circumstance.

In 2010, a graduate nursing student created the Women's Group at Central Lutheran Church. One frequent visitor identified the lack of a drop-in center for women in the inner city. Once a week the group meets to discuss any issues of concern, share in creative expressions such as singing songs or writing poetry, and relate over the comforts of food. The women have been active in inviting policy makers and politicians to their group in order to voice their concerns and create wanted change.

Funding at the ACHC is unique. In general, the funding for the ACHC has been private donations from students, faculty, organizations or volunteers. This allows policies and procedures at the ACHC to be fluid and change in response to community input. For example, we are not obligated to follow a strict set of guidelines as written in a grant proposal about the qualifications to receive socks. We do not have to ask people where they are living or why they need the socks; we can simply provide them with socks. There are guidelines in place about the number of socks that are available daily due to funding constraints. Once the set amount of socks is empty for the day, we simply ask people to return the next time we are open to receive socks.

Evaluation follows the principles of developmental evaluation at the ACHC (Patton, 2011). We gauge success through direct feedback from our visitors, whether it is verbal or continued participation at the Health Commons. When people meet one-on-one with the nurse, limited demographic data is collected. The nurses collect data on the following topics: nursing care provided (education, hands-on, counseling, referral, and supplies provided), new/repeat participant, age group, ethnicity, permanent/non-permanent housing, and gender.

This data is then complied into a spreadsheet, which is used to create the statistical data for the annual report. These numbers are viewed as a "glimpse" of whom we see because keeping track is not the focus of the interaction, but numbers give some information to the broader community.

Health Commons in Cedar-Riverside Neighborhood

The Health Commons in Cedar-Riverside came into existence through a series of unique events. A group of local nurses conducted a participatory action research study with Somali immigrant women in the Cedar-Riverside neighborhood of Minneapolis to discover the barriers these women encountered when accessing healthcare (Pavlish, Noor, & Brandt, 2009). The women who participated in focus groups reported a disconnect to the health-care system secondary to lack of relationships found with healthcare providers, the separation of health and spirituality found in the Western medical paradigm, language barriers, and the limited knowledge of cultural norms experienced by providers (Pavlish, Noor, & Brandt, 2010). The research participants requested a place they could visit in the neighborhood, free of charge, to ask health related questions without appointments, interruptions, or timelines.

The authors of the study brainstormed on options to fulfill the need identified by the women. In the fall of 2010, they decided to ask Augsburg College nursing faculty members to collaborate with them to create a Health Commons in a new setting modeled after the Augsburg Central Health Commons. After many extensive discussions and finding space within the neighborhood, the Health Commons of Cedar-Riverside (HCCR) opened its doors to the community in September of 2011. This new drop-in center was not only to be a place of gathering around health and healing, but was to demonstrate collaboration of organizations within the community. The partnership of the Health Commons Cedar-Riverside is among Augsburg College nursing faculty, University of Minnesota Medical Center-Fairview (UMMC), and the East Africa Health Project.

To better understand the population served at the HCCR, it is important to highlight a few demographics. For example, the income level in the Cedar-Riverside neighborhood is the lowest in Minneapolis (City of Minneapolis, 2009c). The majority of visitors at the HCCR are East African immigrants. The well-being of these immigrants, who are mostly from Somalia, is profoundly affected by social determinants of health due to lack of income, education, healthcare access, and adequate living conditions (Helmstetter, Bower, & Egbert, 2010). Immigrants are forced to adapt to a different way of life in a new country where language barriers, social isolation, and lack of knowledge of resources also exist. Additional stressors include experiencing racial disparities, lack of food access, safety concerns, and substance abuse (Helmstetter, Bower, & Egbert, 2010).

The Health Commons: A Handbook of Nursing Practice

The original space for the HCCR was in a room connected to the women's side of a Sufi mosque. The HCCR was originally designed to be a warm, welcoming space for men and women. The space itself was divided by screens to respect the cultural practices of separating men from women when engaging in health practices. The walls were covered with beautiful cloth hung by the women of the mosque. Other bright decorations filled the room that made it feel warm and culturally appropriate.

Since this time the HCCR has expanded throughout the community. The primary location of the Health Commons currently is in Chase Building of the Cedar-Riverside Plazas. These high-rise buildings are historical landmarks in Minneapolis, and are privately owned. The building owner has provided the Health Commons with its own space free of charge, 24 hours a day. This location has programming Monday through Friday. Additionally, the Health Commons rotates its activities between four locations in the neighborhood on Fridays. For example, the first Friday of the month the Health Commons drop-in hours are held at a local mosque and the second Friday are at a community center. During this "mobile" time, massage, healing touch, and one-on-one discussions with a nurse or doctor are available.

There are two main streams of visitors at the HCCR. Many visitors come to meet one-on-one with nurses or volunteers to discuss health related concerns. People may request a blood-pressure check or simple remedies for general pains during this time. Massage, healing touch and Reiki are also offered to people who visit the HCCR during these hours. This is a time to develop trust, make connections and get feedback from community members. As these relationships develop, many people begin to share their story of past trauma, stress or worry for loved ones' well-being, which is often the underlying cause of many physical pains.

The second stream of visitors are those who attend classes, groups or events hosted at the Health Commons. Women from the community have expressed a desire to know more about nutrition, exercise and parenting. Because of this, these group classes were organized and offered to anyone free of charge. Throughout the last two years the Health Commons has offered many classes: nutrition, yoga, strength training, parenting, sewing, knitting, prenatal, gardening, cooking, public speaking, theatre and walking. These group activities have served as a place to promote social connectedness and relationship building amongst participants.

Funding at the HCCR differs from that of ACHC. The collaboration of the HCCR has applied and received grants from such organizations as UCare and Blue Cross Blue Shield. UMMC is the lead on grant management for the Health Commons. Financial support has been provided to implement new programs, provide stipends for bilingual community liaisons (BLCs), and supplies. Bilingual community liaisons are community members who help us make connections in the neighborhood and assist with language barriers.

Having grant funding changes the dynamics and the evaluation methods at this location. For example, data collection must be compiled in order to complete quarterly reporting demonstrating that the outcomes outlined in the grant proposal have been accomplished. This changes the ability to alter or develop programs at times, depending on the grant agreement. In general, most funders have understood the purpose and philosophy of the HCCR, allowing more flexibility in required reporting. For example, the HCCR staff and nurses do not collect names of visitors because this activity could create unwanted barriers and decreases the level of comfort felt by guests. The HCCR funders have understood this practice, and have supported the HCCR in its efforts, without requiring the collection of names or other personal information from guests.

The other evaluation piece that must be taken into consideration at the HCCR is that outcomes and community involvements required are different for each partner. For example, UMMC is a non-profit hospital that is required by the federal government to be involved in the community at a certain level in order to maintain their tax-exempt status. Also, the changes in the Affordable Health Care Act have increased required community outreach efforts needed by such organizations. These requirements change the programs because the UMMC partnership is under different constraints and pressures than the other organizations in the collaboration. This has impacted the work at the Health

Commons because there are many more programs and people involved with the work. Often, decisions will be made or programs will be changed without all partners' input because of the size of the project, which can make the program feel less personable and more medical at times. All collaborators do their best to stay connected and inform each other of changes or events. Also, it means that the level of trust within the partnership itself must be strong. For example, if a change is made, Augsburg College nursing faculty trust that the changes the Fairview Health Systems partner is making is based upon the model and revolves around relationship building.

    The data collected at the HCCR has changed over the years. Often, Masters in Public Health (MPH) students or the Health Commons Americorp VISTA volunteer will compile the data into charts and graphs used primarily for reporting purposes. The primary emphasis of the data collection has been focused on storytelling, which has proven harder to gather because of the time it takes the volunteers, nurses or staff to record the narrative events. Many times when the partners are asked to present on the work at the HCCRs, the emphasis will be on sharing stories throughout the presentation to create a stronger human connection to the HCCR work than simply numbers and graphs.

The Health Commons: A Handbook of Nursing Practice

CHAPTER 3: Rules of Thumb

'Rules of Thumb' in the Augsburg College Doctor of Nursing Practice (DNP) courses are articulated as a means which allows students to share their practice wisdom in simplified, comprehensive terms. Students create 'Rules of Thumb' to describe their mētis, or practical knowledge in local context (Scott, 1998). In this way a nurse can describe the general principles guiding his or her practice to someone outside the health care field. This process minimizes the role of the expert and creates a stronger sense of understanding amongst the community.

#1 Rule of Thumb

Suspend Judgment

Suspending judgment is an important step for nurses, students and volunteers to take when engaging at the Health Commons. Many times people come to the Health Commons with preconceived notions, or judgments, about the populations being served in the space. Everyone is asked to first identify any biases they may bring and to challenge them throughout the day. For example, one student asked, "There was a woman that asked me for a pregnancy test. She already has three children, has no job and no husband. Why would someone in that situation not use birth control?" There is no answer to questions such as these. The situation this woman is in is complex and we cannot begin to understand her everyday reality or struggle. In this instance, we simply encourage the student to suspend his or her judgment and know that we most likely will never know the details or depths or her situation. In that moment, it is our role as nurses to listen, provide support and be present with that individual.

Suspending judgment also means making ethical choices, or "human decisions" as described by Dr. James Orbinski (2008). It is easy and common to label people with a diagnosis or disease, making the situation less personable and over-simplified. Often we do not have any idea what events led to a person's current condition or situation. At the Health Commons, we want people to truly think about that person as a whole. What is his or her story? Does she or he have support? What challenges does this person face daily? What are this person's strengths?

For example, a 21-year-old was recently dismissed from a local transitional shelter because she had intentionally disconnected her insulin pump three times, where each incident ended in a hospitalization. The shelter discharged her because they said she was a liability. What this young woman did not share with the shelter staff was that when she was seven years old, her mother started prostituting her to support her cocaine habit. The authorities learned of the situation when this child was eleven. Her mother was then imprisoned, and the girl was placed in the foster care system. Since that time, this 21-year-old has lived in

86 different homes or treatment facilities. She is developmentally delayed because of the trauma experienced, and she was never able to successfully transition into young adulthood. No one asked her the deeper questions about her decision to remove her insulin pump, which was a form of coping and a cry for help from this young woman. If they had, she may have possibly received the support and counseling she truly is in need of and she wouldn't be exposed to the risks or violations that exist while living on the streets. The only way we are all going to change our world is to be aware of biases we have, challenge our way of thinking, and incorporate our experiences into a new way of understanding.

Activity 1.

=ACTIVITY=

IDENTIFYING BIASES AND SUSPENDING JUDGEMENT

Before entering a community, it is important to identify any biases you may have about the people who live there. For example, a person may believe the homeless people are 'lazy'. Biases are usually learned through our experiences in our community or in social settings. Before entering the community you will be working with, please identify 3 biases. There is no right or wrong and you will not be judged for your thought - it's important to identify are preconceived notions.

1._____

2._____

3._____

Challenge 1.

You are helping distribute socks on a busy Monday morning at the Augsburg Central Health Commons. A woman comes in wearing designer jeans and it is apparent she has a fresh manicure. After asking for socks, the woman inquires about possible free bus passes. She says that she is trying to get to job interviews, but has no way of paying for transportation. She currently is living at a nearby shelter and is receiving general assistance from the county.

Suspending Judgment: Examining the Facts

- What are some personal judgements that come to mind about this woman's situation?
- Have you had any experiences similar to this one? What did you learn?

INSIGHT: Many times people who are living in poverty tend to place high value on their physical appearance. For example, women will say they don't want to appear as though they are living in a shelter because they are feeling extremely judged by society already. Also, people living in poverty rarely get to make choices. Sometimes it's about being able to decide what is most important in that moment.

TRUTH: This woman just left her abusive boyfriend who consistently told her she was 'ugly.' She lost her job because of his repetitive phone calls and appearances at her place of work. She spent the last of her monthly economic assistance money getting her nails done because she was worried about her appearance while trying to interview for a new job as well as she wanted to boost her self-esteem. Her designer jeans came from a free clothing closet and this is her third day in a row wearing them.

The Health Commons: A Handbook of Nursing Practice

The quote below is from a homeless man. People at the ACHC were asked if they would give their dirty socks to the faculty member that day. People put their used socks in all sorts of places throughout the church and would tell her where she could find them. When the faculty member asked this gentleman why people just weren't giving her the socks directly, this was his response.

It's all about pride baby doll... You don't want people to know how you are living. They are protecting their dignity. Sometimes that is all a man has.

#2 Rule of Thumb

Health is Membership

This 'Rule of Thumb' is inspired by Wendell Berry's essay, Health is Membership (2002), which provides insight into the importance of community in order to achieve an optimal state of health. To be isolated in the world can create negative health implications. Our society demonstrates this idea, as books are written about people dying from a broken heart, movies are produced depicting the horrible events of being abandoned on a desert island, and songs are written describing the desperation of a person to feel as though they are loved. The community in which one lives can provide loving relationships, healthy friendships, a healing space, and can allow for a person to find a new level of wholeness and connectedness. This experience is what we are striving to create at the Health Commons- a community that promotes a sense of belonging and an emphasis on relationships.

Nursing theorists have identified the importance of a healthy community in order for health to be obtained. For example, Jean Watson's theory of transpersonal caring involves the human experience, personal and spiritual connectedness, as well as unity with one's universe (Watson, 2005). Madeline Leninger developed the Sunrise Enabler, which involves the need for cultural understanding in order to promote healthy relationships (Leininger & McFarland, 2006). Margaret Newman focused on the patterns of relationships, and the use of intentionality in order to care for the whole person, not just the physical well-being (Picard & Jones, 2005).

When participating at the Health Commons, it is key that belonging and hospitality become the emphasis of actions. People are coming to this space looking for comfort, friendship and healing. If our society continues to separate these entities, ridding ourselves of human experiences and connectedness while disengaging ourselves from one another, the level of health we can achieve will continue to be limited and our local communities will suffer.

Case Study 1.

Case Study: NEIMO

One Wednesday afternoon, an elderly woman was discussing her constant body aches and frequent headaches with one of the nurses at the Health Commons in Cedar-Riverside. The woman, Neimo, was new to the Health Commons and seemed apprehensive to share much personal information with the nurse. Neimo and the nurse discussed the physical conditions and possible alignments at length. Then, the nurse asked, "Do you have family that can help you with doing some simple exercises or would you like some black seed to help?" (Black seed is a traditional healing remedy in the Muslim beliefs). The woman paused for a moment and gathered her thoughts. She then began describing how her sons are back in Somalia, and she hasn't heard from them in quite some time. She said she missed her home country, and being in the U.S. had made her feel alone and unhealthy.

Neimo and the nurse continued to have a long discussion about the current stressors affecting her life, and then suddenly, Neimo stood up and said something to the other women in the room in Somali. Then one woman grabbed a small drum, and the other women gathered in a circle and began clapping and dancing. They all took turns dancing in the traditional Somali style called, Brahma Baba. The women were smiling and laughing as they celebrated their culture, the day and the friendships they had developed with one another. It was a beautiful event that simply unfolded in that moment in time. The physical pains Neimo experienced seemed to suddenly resolve. Over the next few months, the women gathered in that space at that same time to dance together and to teach others the traditional dance.

REVIEWING THE STORY: Health is Membership

- What physical and emotional changes did this woman exhibit throughout the story?

- What created the change in her demeanor? How does this demonstrate the idea of 'Health is Membership'

#3 Rule of Thumb

Hospitality is an Art Form

In order to be welcoming to all people who visit the Health Commons, hospitality must become an art form. It can be as simple as offering someone a cup of tea or coffee. Every shelf, table, or chair must be placed with much thought. Will a table form a physical barrier for guests if it faces this way? Will a medicine cart distract from our conversation? Listening means sitting down, asking reflective or clarifying questions, having a conversation, and learning from one another. There are no agendas or lists of questions that must be answered. It is the development of a trusting relationship. Without listening, we are merely going through the motions without knowing the journey that has brought us both here this day.

The Health Commons: A Handbook of Nursing Practice

Activity 2.

=ACTIVITY=

Try this activity:

1. Place a desk or some sort of physical barrier in the space to separate yourself from those who are entering the room. Record people's reactions and if people seemed apprehensive to enter the space with the barrier in place.

2. Now, move the barrier and intentionally open the space. Now, observe and record people's reactions when entering the space.

For example, look at how the carts with supplies are intentionally placed to eliminate a physical barrier when visitors enter the ACHC. Imagine if the volunteers moved the carts directly in front of them and stood up. How would that change the dynamic? What would make you feel the most welcomed?

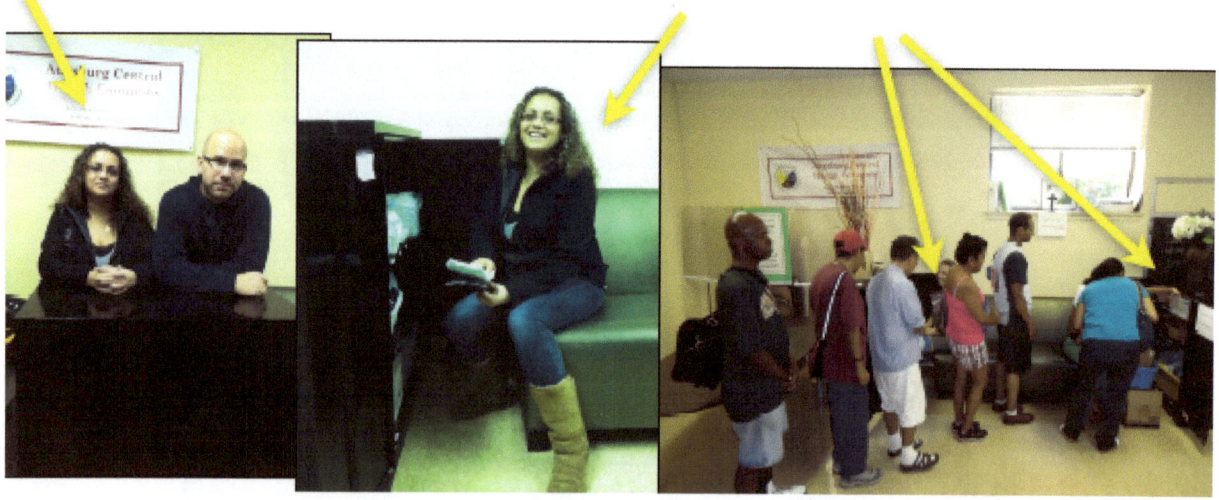

#4 Rule of Thumb

Partnerships Need Care

Partnerships, like any relationship, need attention and care. If you are starting a Health Commons, there will be times where you and those you are partnering with will not see eye-to-eye on a situation, as individual and organizational goals or objectives may cause tensions. Be sure to take time to pay attention to what might be concerning or motivating your partner. If your partnership is suffering, others in the community may start to notice and it can affect your interactions. Take time outside of work to develop a friendship with those you are partnering with. Also, try to envision the pressures or expectations another person may be experiencing. At the Health Commons, it's all about relationships, so pay attention and care for your partners!

Challenge 2.

CHALLENGE

One-on-ones are a civic skill used in community organizing to promote a public culture that is built on understanding people's self-interest and motivation (Boyte, 2008). This is an excellent tool to use during time of conflict or when simply building your partnership relationship. Here is an excerpt from the book, *The Citizen Solution*, by Harry Boyte, discussing one-on-ones:

"A one-on-one involves a conscious exploration of another person's interests, passions, most important relationships, and stories. One-on-ones depend on putting aside prejudgments and stereotypes and listening carefully and strategically...One-on-one interviews are also a way to develop new power through building public relationships across lines of difference.. (they) aim at 'public knowledge'-you are listening for people's public interests and potential to take action with others." (p.32-33)

A TALE OF A ONE-ON-ONE:

During a leadership meeting, two of the partners were disagreeing over the data that nurses should track on visitors from the Health Commons. Throughout emails and further meetings, the two continued to lack understanding of each other's perspective. One of the partners decided to do a one-on-one with the other partner. This event changed the dynamic entirely! The partner being interviewed discussed the lived pressures she had from her organization to collect certain information that involved multiple layers including governmental requirements. In the conversation, the interviewee stated she assumed the interviewer was aware of this situation, which she was not. The one-on-one provided time for the interviewer to listen and reflect on what the interviewee was sharing. Now, this issue is better understood in the partnership and the value of one-on-ones was exhibited.

Examining the Process: ONE-ON-ONES

1. Be prepared: Set up the interview in advanced, think about what you want to know and keep it short (30 minutes)

2. Keep it informal: Go with the flow and pay attention to body language.

3. Look for connections: Find similar or contrasting experiences, but remember, the interviewee should be doing most of the talking!

4. Ask direct questions: What public issues anger or energize the interviewee.

5. Avoid asking yes and no questions: If you do ask them, follow-up with "Why?"

6. LISTEN well: Build on what the interviewee has already said.

7. Be sure you understand: Clarify what the talker is saying by restating what you've heard.

8. Look for energy and action: If you can see that the person is fired up about a public problem, ask if she/he has ever taken action on it before and how.

9. Evaluate: Reflect in your journal after the one-on-one and process the conversation.

   -(Boyte, 2002, p.33-34)

#5 Rule of Thumb

If It's Easy- You're Not Thinking

It is easy to get caught up in the tasks of the day at the Health Commons and to forget to reflect on the actual significance of the events. For example, a nurse could simply focus on stocking the cart, providing basic supplies, making eye contact, getting through the day, and going home. But, this task-orientated mindset hinders the ability of people to develop trusting and meaningful relationships at the Health Commons, which is vital to our purpose. We need to focus on learning about someone's way of life, hearing someone else's story, understanding the circumstances a person faces, and promoting self-discovery. Nurses, volunteers and students should be questioning the day's events, thinking of the societal influences on people's lives, and asking oneself, "What role can I play in creating needed change in my communities, my neighborhoods and my world? How will my experiences at the Health Commons change my professional or personal way of being? What is the learning I can take away from this day?"

Activity 3.

TAKING FIELD NOTES + JOURNALING

Taking field notes and journaling are excellent ways to reflect and gain understanding in your work and interactions. Field notes are observations or conversations that you write verbatim directly after or during an experience. They are usually more of a 'raw' form of note taking. Journaling is your reflections on what these experiences meant to you. When you gather data in this way, you are able to process the activities and the interactions of the day in a more meaningful way.

=Activity=

Everyday you are working in the community or at the Health Commons, take 30 minutes to reflect by either recording field notes or journaling. At the end of one month, review your data and reflect on the following questions
1. What did I experience demonstrating "Health is Membership"?
2. What realities did I learn about people living in poverty?
3. Did a relationship change or develop throughout the last month and what was significant about that experience?

The Health Commons: A Handbook of Nursing Practice

#6 Rule of Thumb

Being Truly Present

Being truly present may sound simple, but it requires a person to practice intentionality and the art of *being* with another human.

Before each day, it is important to take a **DEEP BREATH** and **attempt to rid oneself of outside stresses and distractions**. This encounter might be the one chance a visitor may get this day, this week, or even this month to be greeted with a **SMILE** and ask for something without being questioned about personal information or how the item will be used. If nurses, volunteers or students do not work on doing this every day, people may be distracted from truly listening in conversations or fully welcoming people who enter our doors. It simply means **letting go of outside distractions and engaging in relationship** with all participants at the Health Commons. It is to **BE PRESENT IN OUR WORLD**, the here and now, because this is what is most important at this time.

#7 Rule of Thumb

Mutual Benefit

Mutual benefit occurs when both individuals in relationship gain from a friendship/circumstance. For example, when students first came to the Health Commons, it was assumed that they were providing a beneficial service to people living on the streets. Through the experiences, it became apparent that the students too where benefiting as they were able to gain understanding into the culture of poverty, an often hidden culture in our society. Through the work at the Health Commons, both students and visitors gain valuable knowledge, experiences, and insights from the relationship.

Incorporating mutual benefit throughout the happenings at the Health Commons means that when a new or current program, event or activity is organized, it must be examined for its potential mutual benefit. For example, if someone's self interest is to start a yoga group in order to fulfill required practica hours, but the community members have not expressed a desire for participating in yoga, there is no mutual benefit. If a group of women from the community expressed an interest in beginning a yoga practice, and the student desired to learn about the culture of the neighborhood through consistent engagement and relationship building, then the activity is about mutual benefit. Finding out people's self-interest in participating in the Health Commons is a very important part of respecting the practice of mutual benefit.

#8 Rule of Thumb

Laughter is a crucial part of human interaction. There is a time and place for humor, but over the course of a relationship, it is important to find joy in the relationship. It means playing with a child, asking someone about his/her favorite sports team, learning about each other's families, and sharing our lives with one another. Difficult situations will arise, and sometimes people need to cry, sometimes they need to share, and sometimes they just need to laugh. Whether women are dancing and creating poems in the Women's Group, or a regular visitor shares a joke; it is about a relationship being whole. It is about finding joy together.

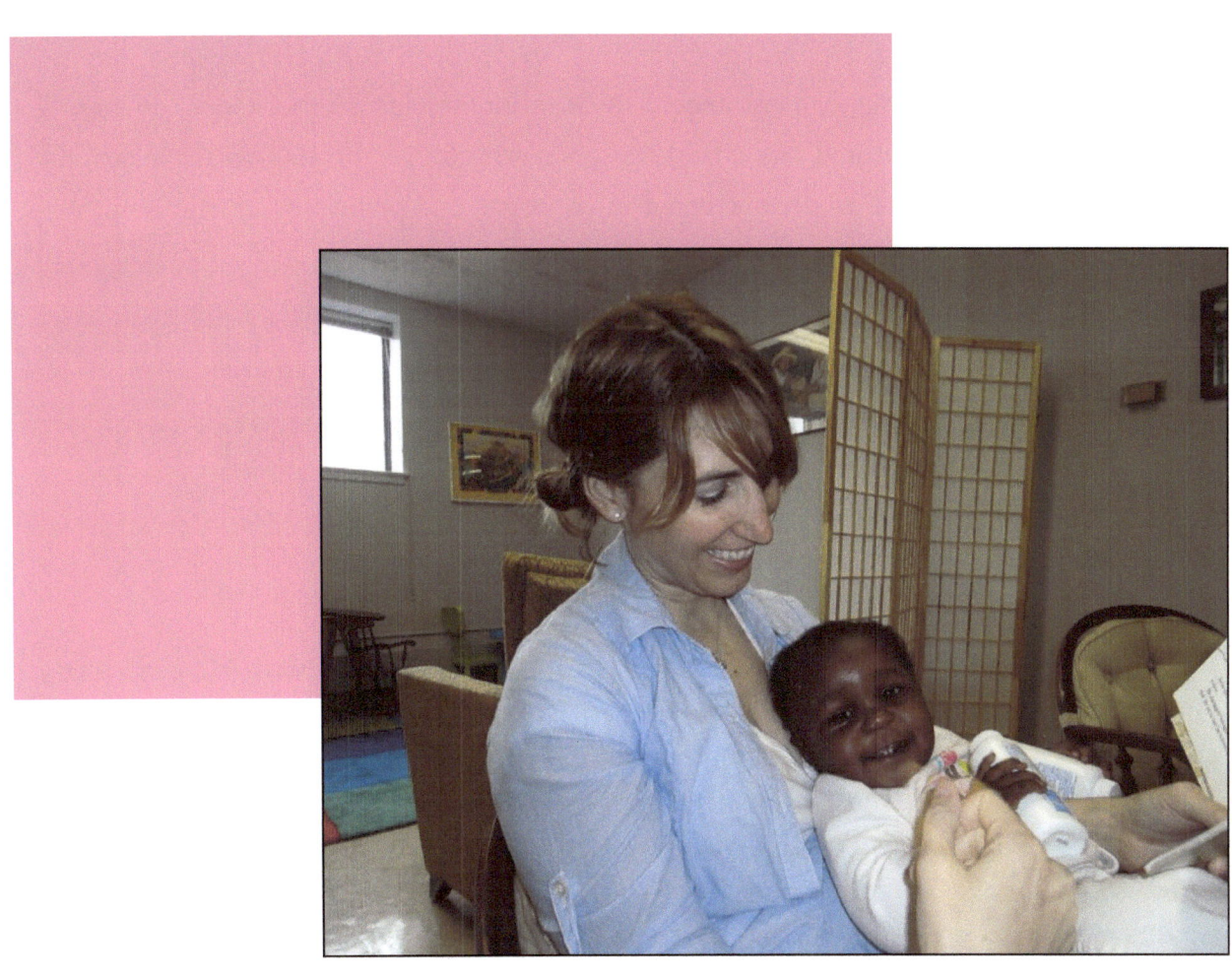

The Health Commons: A Handbook of Nursing Practice

CHAPTER FOUR: The Stories

Using the art of storytelling to give voice to those with whom we interact is crucial to the work at the Health Commons. We prefer using stories as a means of sharing our successes or challenges. By sharing stories of the impact human connectedness created and the situations encountered at the Health Commons, people are better able to understand the concepts and principles of the Health Commons.

Story 1.

Randall:
application of the model

One Friday morning, Randall was at home in his apartment with his girlfriend and three-year old child. Randall, a twenty-two-year-old man, had not finished high school because he was eager to get into the work force to support his family. One afternoon, as he went to look for his three-year old who had gone to the neighbor's apartment to play, he walked in on his neighbor, a grown man, molesting his daughter. Filled with rage, Randall fled to his apartment to find a weapon. He grabbed a knife and went back to the apartment, where he stabbed the neighbor to death.

Charged with second-degree murder, Randall spent seventeen years in prison. During his time there, he learned to read and finished his GED. His girlfriend had moved on to marry another man and started a new family. Once released from prison, Randall found himself a job doing construction and began to make a life for himself.

Randall then met Tonia, a young, energetic woman who was a waitress at a local diner. Soon, they married and she was with child. Tonia was eager to move and start fresh in a new state. The couple decided to move to Minnesota. Once in Minnesota, Randall found himself a full-time job, an apartment and a new journey began. Soon after their arrival to Minnesota, Randall's second daughter was born.

Randall came home from work one evening to a deserted house. Much of the apartment had been emptied and his wife and daughter were nowhere to be found. Panicked, Randall ran to the police to file a missing persons report. The police told Randall he must wait 48 hours to file a missing persons report, and to go home and wait. Randall was able to meet with a social worker for a few hours to discuss the situation, and then he made his way home.

The next morning, the police came to Randall's apartment with a warrant for his arrest, charging him with domestic abuse. Randall was baffled by the claims. He soon discovered his wife had left him to get back together with her old boyfriend. Actually, the wife's intent to marry Randall and move to Minnesota was to steal his money for her and her previous boyfriend. In the police report, Randall's wife claimed that he hit her in the head several times during an argument. During the court appearance, Tonia claimed this event occurred at the time that Randall happened to have been meeting with the social worker, which dismissed the case. Heart-broken and discouraged, Randall left to go home to his apartment.

A few months later, Tonia showed up on Randall's doorstep begging for his forgiveness claiming that her old boyfriend forced her to do what she did or he would've hurt her. Randall decided to give her a second chance. He longed to be with his family again. Tonia and their daughter moved back in and life seemed back to normal.

The following month, the same situation repeated itself, except his time Randall's bank account had been drained. All $40,000 in savings disappeared. Randall was irate. He thought, how could she have done this again, and now left him with nothing! Randall was not going to go down without a fight. He completed a theft report with the police and filed to get full custody of his daughter. The theft report was dismissed since Randall and Tonia were married, and her name was on the bank account. At the custody hearing, Randall learned the most horrific news of all; the paternity test showed he was not his daughter's biological father. Randall fell into a deep depression. He turned his sorrows to the bottle and marijuana. Soon, he was jobless, houseless and alone on the streets.

Randall's story doesn't stop here. He struggled with addictions and homelessness for years. His marijuana addiction turned to crack cocaine. He was living in and out of shelter in a haze of hopelessness. Randall soon realized he was going to end up dead if he didn't make changes in his life. He found a few local churches that offered free services as a place to seek refuge. One of the churches had a nursing-led drop-in center, called the Health Commons.

He spent many years visiting the Health Commons. At first he came for basic toiletry items. Randall says that everywhere we went he felt judged and isolated because of his situation and his prior conviction, but he was always greeted warmly at the Health Commons. After awhile he started sharing his struggles with the nurses while getting his blood pressure checked. As the visits continued, he started staying longer and sharing more of his personal struggles with the nurses. He found he had a strong personal connection with one nurse in particular. She always remembered his name and his situation. He said, "She would consistently tell me how strong and resilient I was. It really allowed me to start believing in myself." He claims that without this support he doesn't know if he'd ever had the strength to make the changes in his life he so desired.

Randall has now turned his life around. He is inspired to work towards greater societal change involving homeless issues. He began providing tours to anyone who was interested, where he would bring people to visit parks, shelters and drop-in centers to provide a glimpse of the reality of living without a home. He has now finished college and continues working to teach people around the state about the realities of being homeless. He credits the Health Commons as a place that helped save his life. He continues to visit the Health Commons when he is in need of support and works with students on understanding issues of poverty and homelessness.

Application 1.

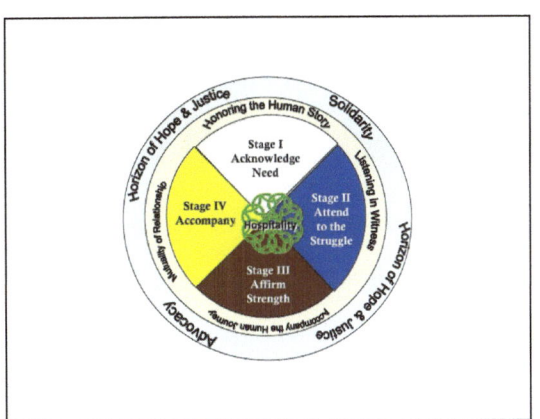

This activity will allow you to analyze the story of Randall and apply parts of the story to the practice model. This process will permit you to gain further insight into the relevance and principles of the model as you move forward.

1.) What about this story stood out to you? Did it influence any of your personal biases?

2.) Did anything from this story represent STAGE 1- ACKNOWLEDGE THE NEED? Please refer back to the section discussing the model of practice.

3.) What from Randall's story represents STAGE 2- ATTEND THE STRUGGLE?

The Health Commons: A Handbook of Nursing Practice

4.) What part of the relationship with the nurse represented STAGE 3- AFFRIM THE STRENGTH?

5.) The final stage of the model is ACCOMPANY. What represents this final stage and continuous stage of Randall's story?

Story 2.

Alexis:
Understanding the Complexity of Circumstance

Alexis was a thirteen-year-old girl from Northern Minnesota. She loved spending time with her friends and received good grades in school. Her main challenge was that she could not get along with her stepfather. One day after school, her stepfather took away Alexis's phone because she did not complete her chores the day before. Alexis was furious! How did this man think he could take away her phone when he wasn't even her biological father! Angered, she used her remaining baby-sitting earnings to purchase a ticket on the bus to the cities for the next morning. She was done being treated this way.

The bus dropped Alexis off downtown Minneapolis near Block E. Reality now set in. It was a Friday morning, and the streets of downtown were bustling with people. She only had her backpack. She had no phone and knew no one. A young couple approached Alexis saying, "You look lost. Can we buy you some lunch?" Alexis, hungry and scared, agreed to accompany the couple to lunch. The couple seemed nice and reminded her of neighbors back at home.

During lunch, the couple asked a favor of Alexis. They asked, "Hey, if we buy you a new outfit, would you be willing to go on a date with our friend? He is lonesome and just wants some company. It wouldn't have to really be a date, but just a way for him to not feel so alone." Alexis agreed to the offer, thinking it would mean not only a new outfit, but also another meal.

That evening, the man, who was Alexis's first john, picked her up from downtown. He was a cab driver who lived in the suburbs. She had no idea what had happened in that cab. She just knew she wanted to go home. When the couple picked her up after her "date" that night, they warned her that if she went home she would never be accepted in her community again. The couple told her that once people found out what she had done, parents would not allow their children to hang out with Alexis because of her reputation and kids would call her a 'slut.' Distraught and confused, she went to bed on the couple's couch. In the morning, she would be pressured again to go on a "date" with more of the couple's friends. The relationship between Alexis and the couple would soon turn abusive. Alexis would turn to drugs to deal with the pain and constant yearning to return home.

Now, fast-forward ten years. Alexis, now with a two-year old, visits the Health Commons looking for resources for shelters for her and her daughter to stay. She has been charged with multiple prostitution offenses. Her parents had found her six months after she had originally come to Minneapolis and she returned home with them for a short period of time only to find out what the couple had warned her about was true. She endured intense psychological trauma because of her experiences with sexual exploitation that caused her to experience depression, insomnia and low self-worth. She couldn't handle the disappointment from her parents when they learned the truth of what she had been involved with or the rumors that spread about her throughout her small town. She came back to the cities after a few short months at home to be with her "boyfriend", a thirty-two year old "businessman". Her life of survival sex, emotional abuse, and drug use continued until the birth of her daughter. Now, she is trying to survive while being homeless.

Application 2.

Alexis appears angry and emotionally withdrawn when she first enters the Health Commons. It is apparent that she has been hurt and lacks trust in others.

Question 1: What would be an appropriate first step to approach this young woman and her child?

1. _____
2. _____
3. _____

Question 2: What are some expectations that you have from the interaction?

1. _____
2. _____
3. _____

Providing Insight: Principles in Practice

Alexis's situation is complex, yet not completely unique to those that visit the Health Commons. In reality, you wouldn't have the insight prior to meeting this young woman and her child. She would simply appear as an angry woman and some might label her as "feeling entitled" to what is available at the Health Commons. There are 3 important topics to highlight about the situation.

Often in the health care setting, there is an urgency to "fix" the situation. Nurses must get away from this mentality when working at the Health Commons. These are complex situations that cannot be over simplified. Before knowing Alexis's situation, the notion that "There is Always More to It" must be applied. We cannot judge her for her angered appearance.

Relationships take time to develop, and Alexis may never trust someone enough to share her struggles. All a person can do be there for her to support her as she identifies a need. Hopefully a relationship will build overtime.

Story 3.

Ayaan:
Moving Away from the 'Expert Model'

One Friday afternoon, after receiving a free massage, one elderly woman suddenly jumped off of the massage chair, and went into a yoga pose on the ground called 'Downward Dog.' The woman was smiling and talking vibrantly in Somali. Through the bilingual community liaison, the nurse providing the massage soon learned that this woman was sharing the benefits she was experiencing from programs at the Health Commons with the nurse.

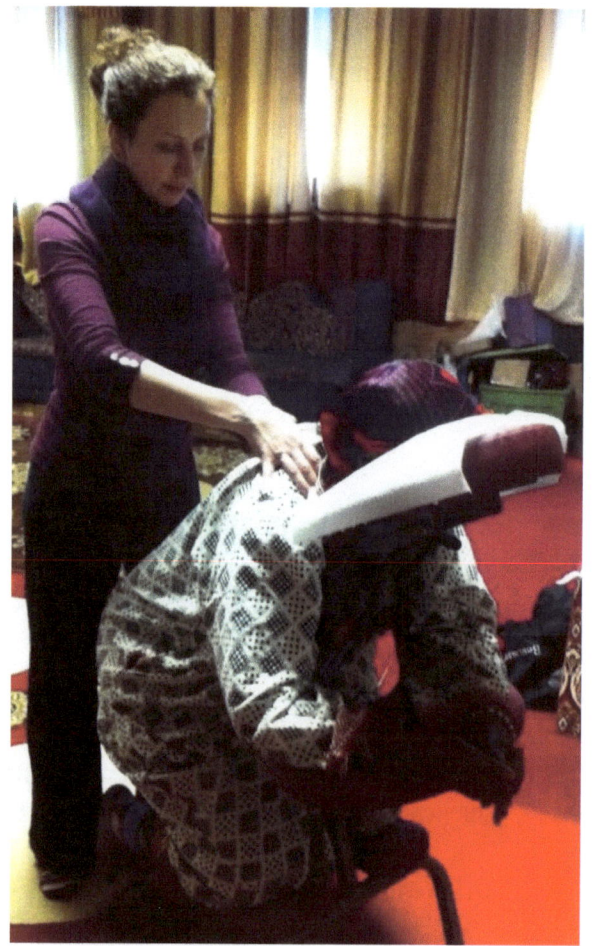

The woman said three months ago, when she first began coming to the Health Commons, she often experienced muscle stiffness and soreness. It was affecting her mood and ability to move around. She was demonstrating to the nurse her ability to move, which she credited to yoga and massage at the Health Commons. She was very happy to be feeling physically fit. This woman then decided that she wanted to volunteer at the Health Commons because she wanted others in the community to be able to benefit from the programs and relationships that the Health Commons offers. She continues to volunteer to this day.

Application 3.

Applying Concepts

Question 1: What stages of the model can be applied to Ayaan's story?

Question 2: Part of the purpose of the Health Commons is to get away from the 'expert model' and allow all people in the space to share in supporting one another and answering health related questions. For example, Ayaan is not a trained health professional, but she wants to volunteer and give back to the Health Commons. She says "It gives me a sense of purpose." She is currently participating in the 'Train the Trainer' program for yoga.

- How do you feel about this concept on getting away from the 'expert model'?
- Do you see value in 'Train the Trainer' programs and volunteerism?
- What does this story demonstrate about the importance on ongoing relationships?

CHAPTER 5: Where to Begin

Starting a Health Commons doesn't involve an objective list of tasks to complete or surveys to be done. What it does involve is being connected to the community and the needs members identify. For example, the ACHC started because homeless visitors at the church were asking about blood pressures. Overtime, the ACHC transformed into a space that continues to help with simple health related questions, but is working on much larger societal issues. For example, the ACHC serves as a bridge between the traditional medical world and those who are experiencing poverty in Minneapolis. The ACHC can help with navigating the system, identifying resources and understanding our complex medical system. In turn, students, volunteers and nurses at the ACHC can help advocate for those we work with. The HCCR was started based on feedback from immigrant women of the community who requested a a safe, easily accessible space to ask health related questions, free of charge without time constraints. The programs, locations and hours are constantly changing and adapting to the needs expressed by community members.

Here are some simple tips for beginning:

Start Small- The more people involved with the initial planning of the Health Commons, the more complicated the process will become. Also, you may become overwhelmed if you overcommit yourself. Start with a few hours a week and expand from there.

Get to Know the Community in Non-Traditional Ways- Are there options such as 'Day in the Life' tours or any informal guides in the local area that can help you gain understanding into the everyday realities of the setting. Explore neighborhood parks or hangouts. Find out what issues community members gather around.

Who Identified the Need and Why?- What reasons do you have to start a Health Commons in the neighborhood? Do members of the community seem supportive and interested in such a center? Why are you or other collaborators interested in starting a Health Commons? Identify the self-interests of the group members.

How Much Funding Do You Need?- You can start a Health Commons with a simple blood pressure cuff and stethoscope or you can get a grant to provide more services. But, remember, it is about relationship building and collaboration- not simply providing a service or a charity. If you receive funding, you have to think about the ability you will have to change programs and what you will be responsible for reporting. For example, do the reporting guidelines parallel with the philosophy at the Health Commons?

It Takes TIME!- The Health Commons is a place where relationships and trust need to build overtime. People may be hesitant to participate at the Health Commons at first. Some people may fear that you are collecting person information. Others may feel as though you are receiving money to work with them. People living in poverty have often been exploited or mistreated by providers, and it may take time for people to build a trusting relationship with you. Time can be a few days, a few months or even a few years!

Get To Know Other Services In the Area- It's important to get to know other non-profits and government agencies in the area. It will help you be more knowledgeable of what is available as well as you may be able to collaborate together in the future. Often, having someone well connected and well respected in the neighborhood help you begin can make all the difference in the world.

Continue to Review the 'Rules of Thumb'- It is easy to lose site of what our purpose is and why we are doing things the way we do them. Going back and reviewing the 'Rules of Thumb' will allow you to stay focused as well as to reflect back on the process in a more meaningful way.

Hold Back On Defining- When you first begin, you may want to define your work quickly and set up a schedule, but it's important to stay open to feed back from the community on what programs or schedules work for them and then be able to adjust accordingly. Don't fear living in ambiguity. The quicker you define your work, the less flexibility you will have for change in the future.

Worksheet 1.

Community Needs

Write down everything YOU think that the community needs? What are the health related issues? How should should these needs be addressed?

Once you have completed this exercise, please put this list away. It is important to identify our preconceived notions about the neighborhood's weaknesses and needs before we learn about it first hand from the community members what is actually needed at the grassroots level and what are the strengths of those who live in the community. You can save this list for self-reflection to be reviewed at a later date.

Worksheet 2.

Informal Knowledge

Question 1: **Do you know people who live in the neighborhood? Can you do a one-on-one with them?**

_____
_____
_____
_____

Question 2: **What local gathering places are there? Go there and learn about it.**

_____
_____
_____
_____

Question 3: **Where are people getting their information? Is it word of mouth? Websites? Find out.**

_____
_____
_____

Question 4: **What issues matter to the community? Attend a local meeting or event.**

_____
_____
_____
_____

The Health Commons: A Handbook of Nursing Practice

Worksheet 3.

RESOURCES

In Hennepin County there are many great community resources to connect with the community, organizations and groups. Please look in the resources below if you are in Minneapolis. If you are not in the Minneapolis area, then please search the internet for possible resources similar to these.

- Call 211 - A free call to the United Way
- Participate in 'A Day in the Life' Tour through St Stephens Outreach
- Access a pamphlet called 'The Handbook of the Streets' which lists all the community resources for those in need
- Visit your local county Human Services and Public Health Department. Ask about resources or processes involved with applying for benefits. Hang out and observe what happens in that space.
- Visit local alliances, schools, organizations and faith-based organizations to find out what they do and learn their insight into the neighborhood. Are they in need of volunteers that would help you start having a presence in the area?

Worksheet 4.

Partners

If you are going to partner with other individuals or organizations, important brainstorming should be completed together. Please read the following questions and answer them to the best of each potential partner's ability.

What is each potential partner's self-interest in joining the partnership?

What required documentation or data collection will be needed from your organization? Are there certain pressures that influence your organization's involvement in the community of interest? For example, grant requirements or policies?

What style of communication does the lead of each organization prefer? Emails, face-to-face, phone conferences, etc.

How much time or money can each partner contribute? Discuss roles and responsibilities for each partner. Write these down and come to a collective agreement on the assigned roles and responsibilities.

Worksheet 5.

FUNDING — Funding is a major issue when organizing a Health Commons. It should be well thought out and assessed prior to creating a new Health Commons.

What do you actually need funding for? If the supplies are really a means to build a relationship, then what are the bare minimums that you will need?

Can you find space for free that will eliminate the need for rent? Do you need translators?

Is there a community center that has space you could use that already has security or other necessary provisions?

If there was no funding, what would happen? Would you and your partners continue or is the project funding-dependent?

Are there grants that fit into your Health Common's goals and vision? Who will manage the grant money if you are a recipient?

Worksheet 6.

When in Doubt: JOURNAL, JOURNAL, JOURNAL!!

Journaling will help you process, reflect and gain insight into the Health Commons. There are many issues and topics that will arise, and this will allow you to intentionally reflect and grow. Also, it is a great way to review the 'Rules of Thumb' or apply the practice model to the new center.

CHAPTER SIX: Words to Live By

In this chapter, a simple list of profound and meaningful books, essays and videos are listed as suggested resources for those who are interested in learning more about the unique model of practice at the Health Commons. Each resource is valuable for its insights and perspectives.

- *An Imperfect Offering: Humanitarian Action for the 21st Century* by Dr. James Orbinski (2008). This book has also been created into a documentary called Triage: Dr. Orbinski's Humanitarian Dilemma

- *Developmental Evaluation: Applying Complexity Concepts to Enhance Innovation and Use* by Michael Quinn Patton (2011)

- *Grassroots Post-Modernism: Remaking the Soil of Culture* by Gustavo Esteva (1998)

- *Pathologies of Power: Health, Human Rights, and the New War on the Poor* by Dr. Paul Farmer (2005)

- *The Art of the Commonplace: The Agrarian Essays of Wendell Berry* (2002)

- *The Citizen Solution: How You Can Make A Difference* by Harry Boyte (2008)

- *The Soloist* film directed by Joe Wright, and starring Jamie Foxx and Robert Downey Jr. (2009)

CHAPTER SEVEN: References

Berry, W. (2002). *The art of the commonplace: The agrarian essays of Wendell Berry*. Berkeley, CA: Counterpoint Press.

Boyte, H. (2008). *The citizen solution: How you can make a difference*. Saint Paul, MN: Minnesota Historical Society Press.

City of Minneapolis. (2009c). Cedar-Riverside Income - Minneapolis Neighborhood Profile. Retrieved May 10, 2012, from http://www.ci.minneapolis.mn.us/neighborhoods/cedarriverside_income.asp#TopOfPage.

Enestvedt, R., McHale, K., Miller, J., Loushin, S., Gunderson, J., Kinney, M.A., Freborg, K., Schuhmacher, D., & Baumgartner, K. (2009). [PowerPoint slides]. *Developing a nursing practice model in the social margins: Community-building as remaking the soil of cultures*. Presentation at Augsburg College, Minneapolis, MN.

Helmstetter, C., Brower, S. & Egbert, A. (2010). *The Unequal Distribution of Health in the Twin Cities*. Wilder Research: Saint Paul, MN.

Leninger, M. & McFarland, M. (2006). *Culture Care Diversity and Universality: A Worldview Nursing Theory*. Sudbury, MA: Jones and Bartlett Publishers.

Orbinski, J. (2008). *An imperfect offering: humanitarian action for the 21st century*. New York, NY: Walker and Company.

Patton, M.Q. (2011). Developmental evaluation: Applying complexity concepts to enhance innovation and use. New York, NY: The Guilford Press.

Pavlish, C.L., Noor, S., & Brandt, J. (2010). Somali immigrant women and the American health care system: discordant beliefs, divergent expectations, and silent worries. *Social Science and Medicine, 71*, 353-361.

Picard, C. & Jones, D. (2005). *Giving voice to what we know*. Sudbury, MA: Jones and Bartlett.

Scott, J.C. (1998). *Seeing like a state*. New Haven: Yale University Press.

Urban Mapping (2011). Cedar-Riverside neighborhood in Minneapolis, Minnesota. Retrieved from www.city-data.com/neighborhood/Cedar-Riverside-Minneapolis-MN.html.

Watson, J. (2005). *Caring science as sacred science*. Philadelphia, PA: F.A. Davis Company.

CHAPTER EIGHT: Further Information

WEBSITES:

Augsburg Central Health Commons- http://www.augsburg.edu/nursing/health-commons/

Health Commons in Cedar-Riverside- http://www.augsburg.edu/nursing/cedar-riverside/

BOTH SITES ALSO HAVE FACEBOOK PAGES!!

augsburg.edu/nursing

Please feel free to contact Augsburg College's Department of Nursing for any further questions. Thank you for your interest in the Health Commons!

A dome, in reality, can be a great amount of extra labor and materials, (not to mention expense) if not properly planned. It is an exercise in problem solving, not just mathematically, but in other ways such as making conventional things like doors and windows fit in with the structure and function properly, and utilizing the space to maximum benefit. Curved walls and ceilings are somewhat limiting when it comes to breaking up the interior space with 'normal' square or rectangular rooms. The main expense and problem to solve for any dome is the exterior shell or membrane, however. It is both walls and roof and is constantly exposed to weather from all sides and must be designed and fabricated with both the elements and expense in mind, including future maintenance. If improperly constructed, the dome could be an ongoing maintenance problem and a perpetual expense.

3 frequency icosahedron

The geodesic home that I built in the early 1970's consisted of a 36-foot diameter, 3 frequency, 5/8 hemisphere and a 24-foot diameter, 5/8 hemisphere, both covered with stucco, and a conventional rectangular garage. Originally the plan called for them to be covered with neoprene-coated 3/4-inch plywood, but that proved to be too costly. Geodesic math and chord factors were obtained in *Domebook 2*, which described several different domes.

External plywood hub and struts

A 24-foot 1/2 sphere was built first to be used as a greenhouse in order to gain practical experience with such a radical structure, and was an important step. It was framed using 2x4 wooden struts bolted to external 3/4-inch plywood hubs, and sat on a footing of cinder-blocks partially buried in the ground. The skin was made up mostly of translucent fiberglass panels on the south side, and 1/2-inch plywood on the north. There were a few clear as well as blue panels of Plexiglas® on the top. (At that point it became necessary to erect a fence around the property.) The plywood was covered with silicone rubber to waterproof it, which worked well except for one drawback. When cured the silicone rubber tended to have an affinity for dirt, and dust storms in Lubbock are a fact of life. The white rubber took on a dirty brown look, impossible to keep clean. An exterior paint would have been a better coating and a lot less expensive, but there would still be the seams to deal with. The external plywood hub, while suitable for the greenhouse, was not acceptable for the other domes because it lacked both strength and aesthetic appeal.

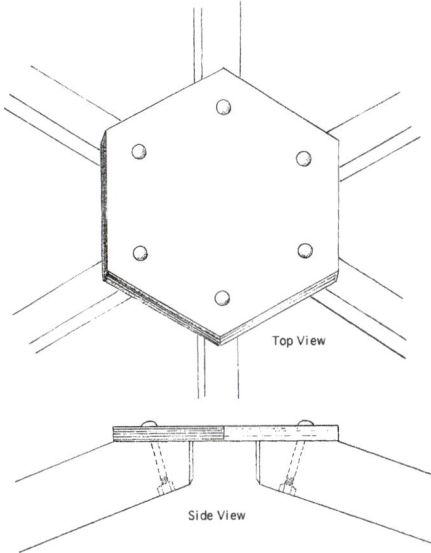
External plywood hub used on greenhouse

Plywood covered with silicone rubber - top panels are acrylic

Internal metal hub and struts

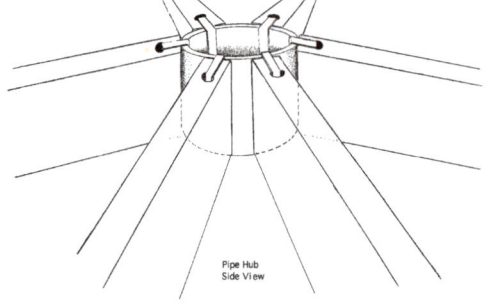

Pipe and strap hub, not used

An internal metal hub which would not show while adding a lot of strength was eventually chosen after a comparative test with the wooden hub. I also built and tested a couple of other fastening systems, including an external metal hub and a pipe and strap hub, neither of which were suitable. One of the potential problems with the internal metal hub system was that it required more precision in cutting the angles of the struts where they all came together. Another issue was that the hubs themselves had to have a high level of precision in their manufacturing. At that time the high pressure water-jet cutting system was not common, nor was the plasma cutting system, so every hub had to be individually hand cut and drilled.

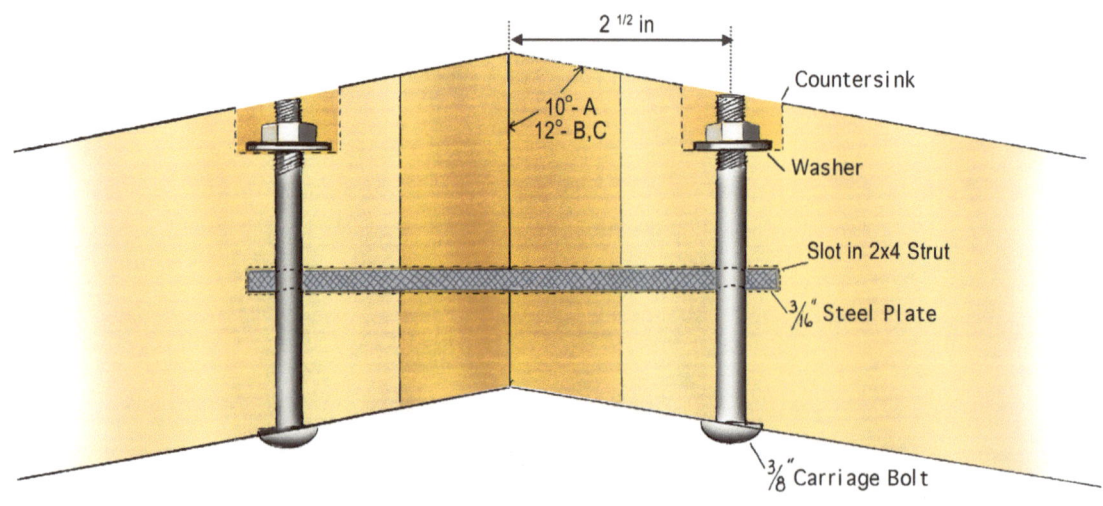

Side rendering of struts and hub

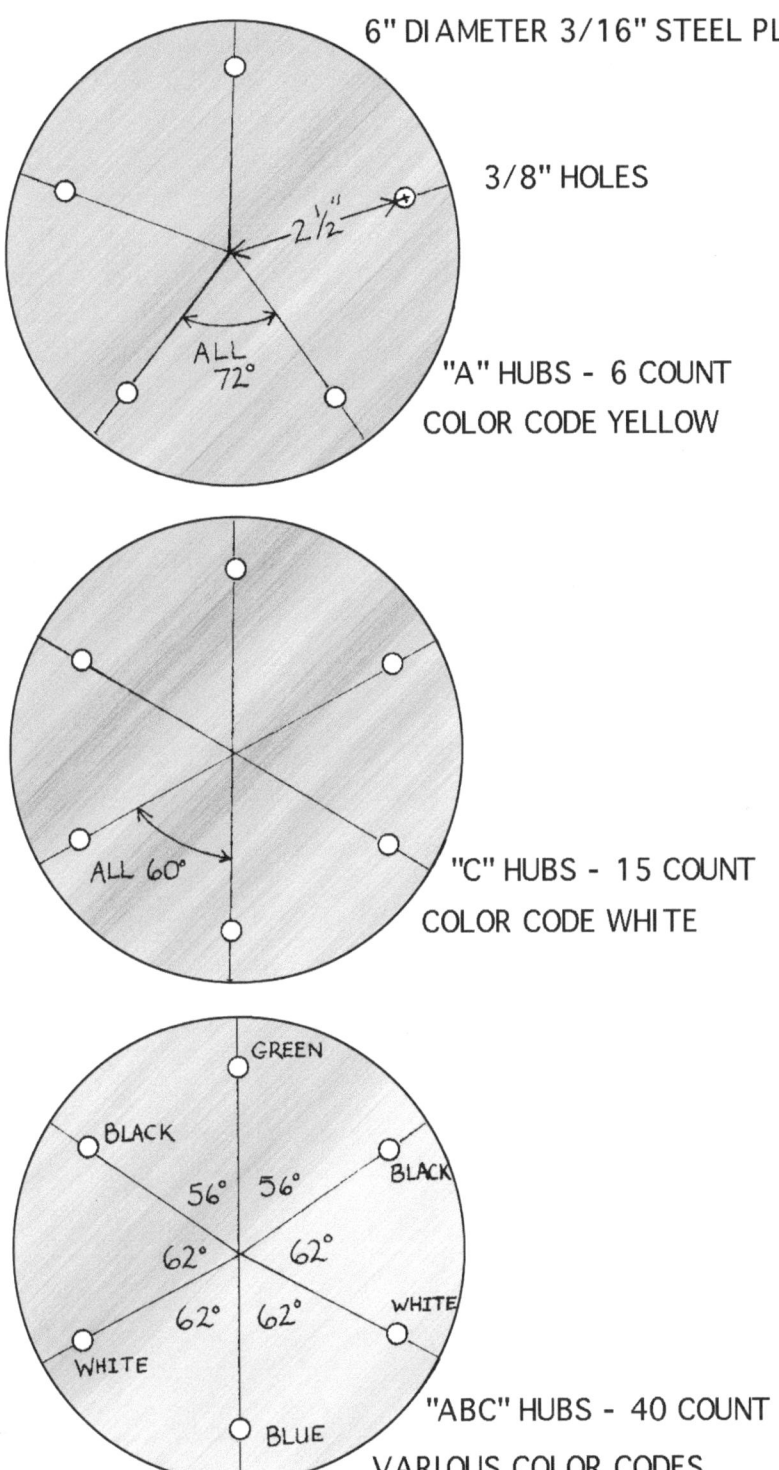

I hand cut 122 six inch diameter circles (hubs) out of a sheet of 3/16-inch thick steel plate, using an oxyacetylene cutting torch hooked to a circle cutting jig. The rough edges were then ground off of each one, creating a smooth disk. Bolt holes at the appropriate angles were carefully marked on each disk using a template, punched to make sure the drill bit went exactly where it was supposed to, then drilled on a drill press and finally color coded. The color coding matched up to the color coding also applied to the struts, absolutely necessary to properly assemble the dome. There were three different hub bolt patterns – a pentagon, a hexagon, and an alternate hexagon. Specifically, there were 6 pentagon hubs, 15 hexagon hubs, and 40 alternate hexagon hubs for each dome.

Note: The ideal way to cut the hubs, including holes, would be the use of a computerized plasma cutter, or water-jet cutter, either of which can be done in an hour with extreme precision. Cutting the hubs, marking and drilling by hand took about 8 to 10 hours.

Next, the lumber for the struts was hand picked in order to obtain the straightest, knot-free, high density wood possible, which is very important. The struts were cut on a radial arm saw to the proper length and angle, using a jig whenever possible, to a tolerance of 1/16 inch. There were 30 A struts, 20 B1 struts, 35 B2 struts, and 80 C struts. The B1 and B2 struts were the same length, but had slightly different angles because the B2s went around the pentagons. The length of the struts was determined by multiplying the radius of the dome by various chord factors; A = .3486, B = .4035, C = .4124.

So the length of an A strut for a 24-foot dome would be the radius of 12 feet (144 in.) x .3486 chord factor, or 50.19 inches. The length of a B strut was 144 inches x .4035 chord factor, or 58.10 inches. The length of a C strut was 144 inches x .4124 chord factor, or 59.38 inches. Longer struts were cut first so that in the event of mistakes the lumber could still be used for shorter struts. 3/16-inch wide slots for the hubs were cut in the ends of each strut, using two saw blades together to equal the thickness of the hub. The next step was to drill the 3/8-inch bolt holes in the ends of the struts and countersink for the nuts, using a jig to ensure proper angling. All struts were then color coded. The careful cutting and color coding of the struts took 1 week.

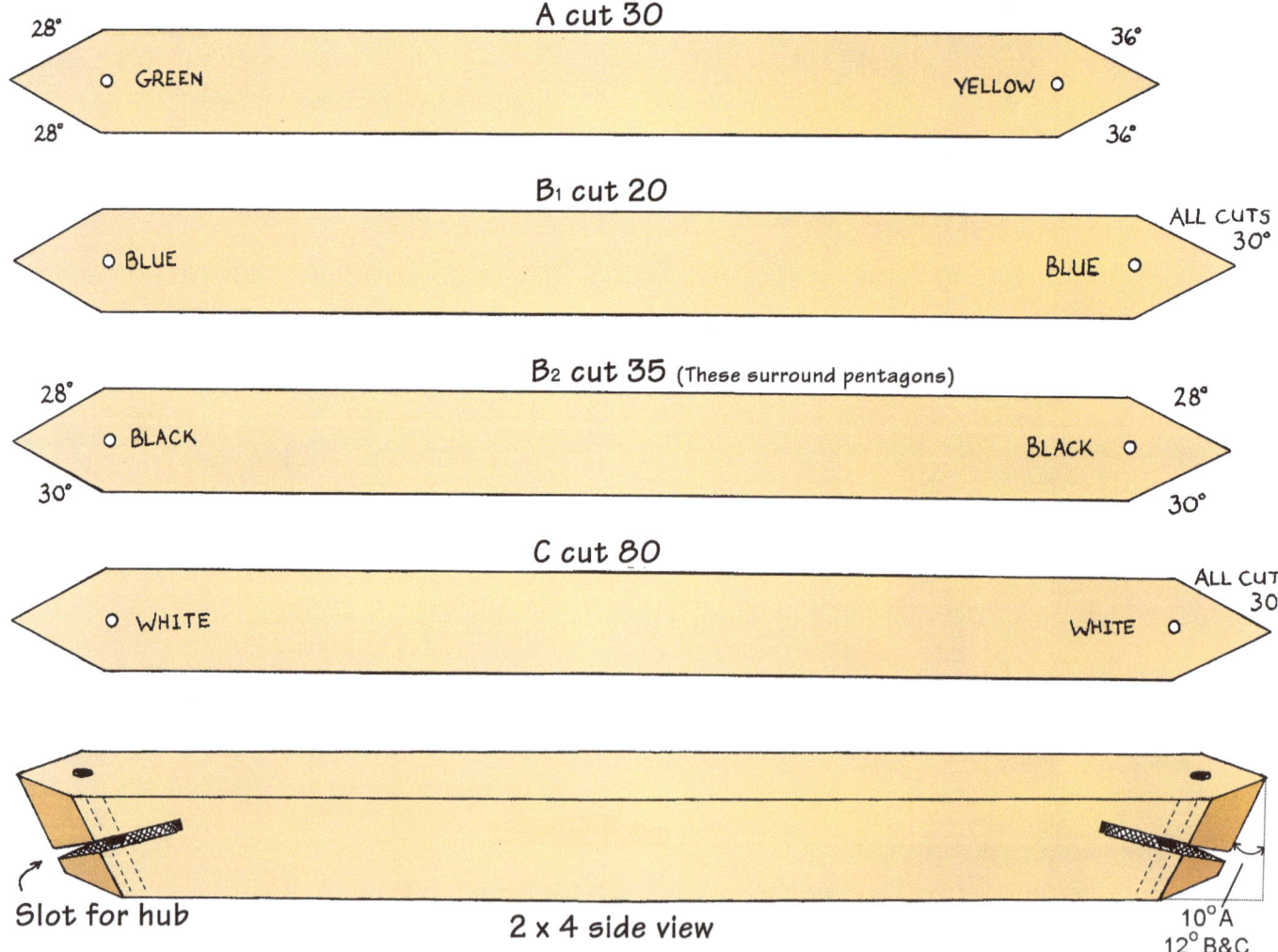

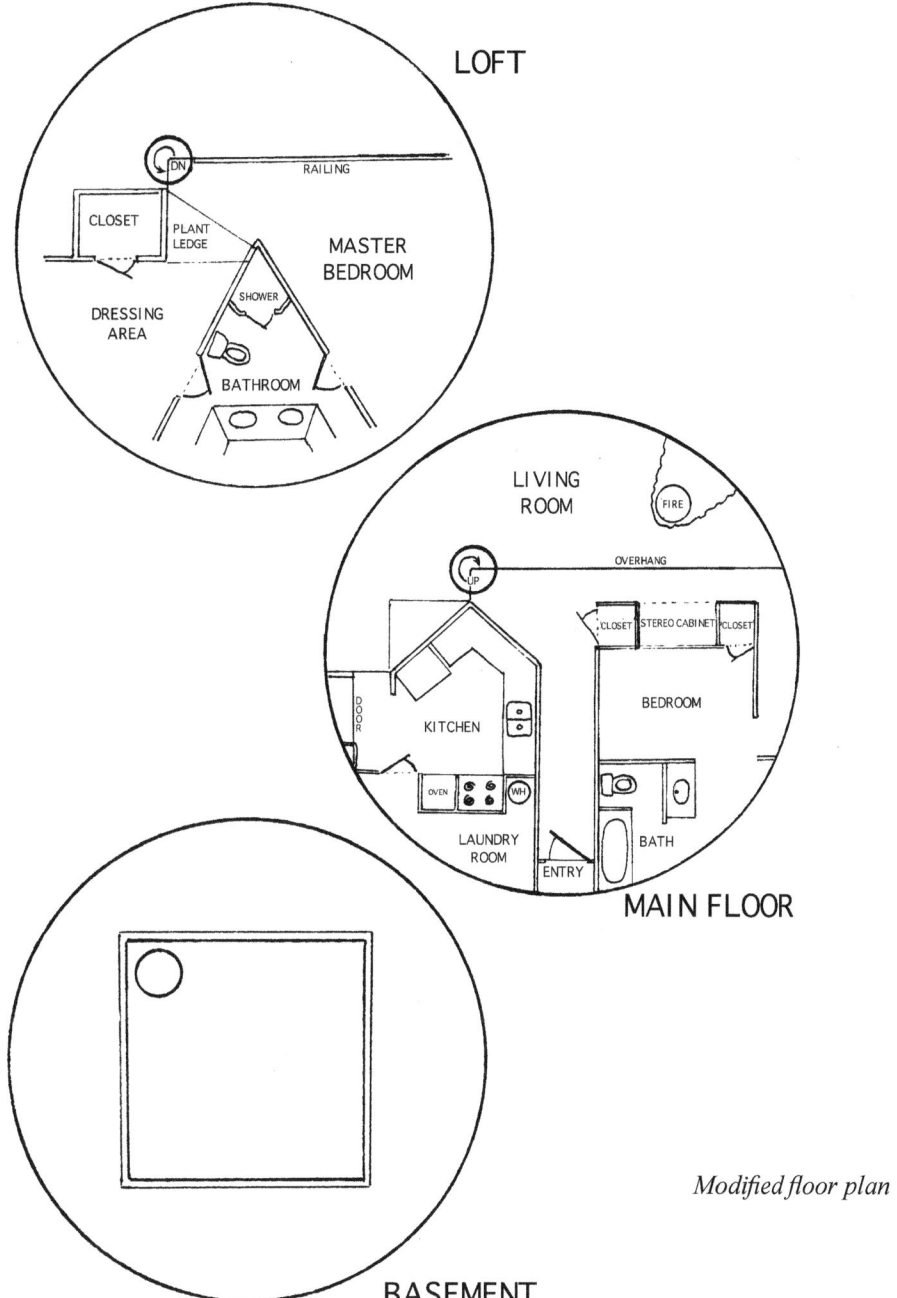

Modified floor plan

Because the domes were to be built inside the city limits of Lubbock, I had to obtain a building permit and a contractor's bond. This meant that blueprints had to be drawn up by a registered engineer and architect, and application made to the building inspection department at city hall. Upon receiving the plans and looking them over, the building inspector's office required further documentation of the load strength of the structure, in spite of the fact that a NASA computer had generated the chord factors published in *Domebook 2*. (Geodesic domes are used by the U.S. Air Force in arctic regions for snow loads and winds in excess of 125 mph!) In order to satisfy their requirements I enlisted the aid of a professor in the engineering department at Texas Tech University. After reviewing the plans he determined that the structure was indeed sound and that we could load test the dome, which would satisfy the building inspection department. The building permit was eventually approved, and various jobs were subcontracted out as required by city codes, like the electrical work. The foundation work was also subbed out, starting with the basement.

Construction had not even begun, and the first deviation from the blueprints was deemed necessary. The architectural firm that I hired had about as much experience as myself when it came to domes – little or none – and produced the blueprints more as a requirement of the building inspection system than an actual working plan. We all knew that there would be problems and that solutions would be found, and that changes were inevitable when it came to actual construction. However, I hadn't planned on making a change before building began, but that's exactly what happened.

Back-hoe digging the basement

The first subcontractor who was hired to do the basement pointed out two major flaws in the basement design. The plans called for a half-moon shaped basement, with the basement walls being directly underneath the exterior dome walls. This was a poor design, as it would allow water to run off the outside of the dome and directly down to the basement wall, creating a potential water leakage problem. Second, the half-moon shape of the basement would have been difficult and expensive to excavate and build using cinder-block, not to mention having a lot of wasted space. He suggested that we build a square basement tucked underneath the center of the house, with its walls well away from the water-shedding exterior dome walls. Heeding his advice, new basement plans were drawn up and submitted to the building inspection office, and once approved we were ready to go.

Basement excavation completed

After determining the footprint of the domes on the site, a square hole approximately 24 feet across was excavated using a back-hoe, and rebar and wire mesh were laid out in the bottom, followed by pouring concrete for the basement floor. There was no plumbing in the basement slab, so the foundation work went quickly.

View from street

After curing for several days the cinder-block walls were built up on the floor, using cinder-block that had a waterproof coating on the outside. The cinder-blocks were supplied by the basement sub-contractor and were from another project he had completed, with the excess blocks being recycled instead of dumped.

Conduit for electrical wiring was run into the hollow cinder-blocks. The hollow cinder-block walls were then filled with concrete and reinforcing steel rods known as re-bar, leaving some of the rebar sticking out the top of the blocks so as to be eventually encased in the slab. The subcontractor and I had some differences of opinion at this point, specifically, during the pouring of the concrete into the cinder-blocks. As I watched the pouring I realized that the concrete was not being tamped with the rebar rods, so I began to do it myself. In spite of his reassurance that it would not be necessary to do this, I knew in fact that it was. Each opening took almost twice as much concrete as it settled, creating a very solid pour and a very solid wall after it cured. He also suggested not back-filling against the wall, but this would have created another set of problems in pouring the slab, so I did it anyway. The lesson here is, when hiring subcontractors, try to be on site when they are doing their work,

Basement walls being built

Basement walls

and get the job description clear with them *before* they do the work. I had several other occasions where this was a problem, partly due to the unusual nature of the project. It wasn't always possible to schedule a sub-contractor at a certain time, so you just had to hope they would come when you were there. It is just as true today as then, but at least cell phones have made this process more efficient.

Rebar embedded in basement walls

Form boards and rebar over the basement

Temporary bracing in the basement

Next, forms were built over the basement which would hold the concrete slab when poured, leaving a 4-foot diameter hole for the spiral staircase. Six 2x12 beams and several sheets of 3/4 - inch plywood used as forms would become a permanent part of the basement ceiling, and were temporarily braced to the basement floor using 2x4s and 2x6s. This was a must, in order to support the weight of tons of wet concrete when the slab was poured. The 2x12s were used to make three channels running the length of the basement, and when poured full of concrete would create three strong beams. A grid of 1/2 to 1-inch thick re-bar rods was wire-tied inside these channels and on top of the plywood forms, creating a sort of web of steel which, once encased in concrete, would give tremendous strength to the slab over the basement.

Rebar tied together

Plumbing stub-outs and curved Masonite® forms

Before the concrete could be poured for the slab, the plumbing had to be roughed in. In order to accurately place the stub-outs and outer form-boards, a center point of the dome was permanently established over the exact center of the basement and a large nail was driven into the forms. All future measurements would be taken from this reference point, which was very important! (Regular houses do not need this reference point as measurements are taken from various areas of the slab.) But no matter what type of structure you are building, double check every sub-contractor's measurements, including your own. I was especially concerned that the next step was done right, measuring for the outer forms for the slab.

The curved outer forms were flexible Masonite®, reinforced every couple of feet by 2x4 stakes. At that point the bottom 15 struts were laid out around the curved forms in order to mark the exact location of the anchor bolts that would be placed

Overall view of the form boards

in the concrete. This was done for both the 24 and 36-foot domes and was a critical step, because the bolts needed to match the location of the hubs within a half-inch. Once the plumbing rough-in and forms were completed, and the rest of the footings, re-bar and wire mesh were done, the building inspector was called in for the first of several inspections. After checking to see that the footings were deep enough,

Cement truck preparing to pour

the steel was in place, and the plumbing rough-in was done to code, a green-tag was taped to the plumbing stub-out, which meant that it all passed inspection and we could go ahead with the concrete pour. (The front entry-way and back patio slabs were done later.) The concrete crew consisted of several people, each working as hard as they could until the pour was finished. The slab was then sprayed with a chemical to keep moisture in the concrete and then the slab was allowed to cure for several days before framing was started.

Concrete being spread

Note: One thing I would do different today is the inclusion of a product called Fibermesh® in the concrete mix. A small bag of this polymer is added to the mix in the truck and allowed to churn for 5 minutes. The result is a concrete mix with thousands of strands of plastic fibers mixed in, which add tremendous strength to the slab.

Concrete being spread

Concrete slab immediately after the pour - note the plywood covering over the basement opening

Concrete slab ready for frame

Framing the domes was very exciting, and went quickly with no problems at all. Color-coded hubs were matched to color-coded struts and bolted together using 3/8-inch carriage bolts and a socket wrench. The hubs that sat on the slab were welded to the bolts previously set in the concrete during the pour. After the bottom level of triangles was completed the structure was strong enough to climb on, and with each level it became more rigid. Scaffolding with wheels expedited the framing tremendously, because on the 36-foot dome it was not possible to climb to the next level on the dome itself.

Framing being started

Assembling the dome

The 36-foot dome was assembled in a total of 2 days, and all of a sudden it attracted a lot of attention. People from the newspaper came by for an interview and took pictures of the dome. Supposedly a small story would be in the Sunday paper in the home section, but I didn't give it much thought. I awoke Sunday morning to the sound of the telephone and a friend asking me if I had seen the newspaper. The picture of the dome and story were on the front page! I had the feeling that I had better get over there. Sure enough, cars were lined up for blocks, and people were climbing all over the dome like a jungle-gym! In order to prevent that from happening in the future, I decided to go ahead and get it covered with plywood, so we focused on that instead of assembling the other smaller dome. From that time on there was a constant stream of curious onlookers with questions and observations about domes, housing and Buckminster Fuller. It made it hard to get any work done at times, but it was interesting to talk with them. On any given day there would be four or five people stopping by.

Article in the Lubbock Avalanche Journal

Framing of the 36-foot dome

Next came the job of sub-framing the big dome's triangles. Sub-framing was the addition of more 2x4s inside each triangle which gave more surface to nail to. This was necessary because the 1/4-inch plywood sheathing used on the exterior was in two sections, (a diagonally cut 4x8 sheet of plywood) and the seam between the sections needed support in order to support the weight of the final masonry coating. By that time we had the 24-foot dome assembled, which took less than a day. The 24-foot dome did not need any sub-framing because the plywood panels were in one piece, able to span the struts with no joints.

Note the sub-framing of 2 inner Ts

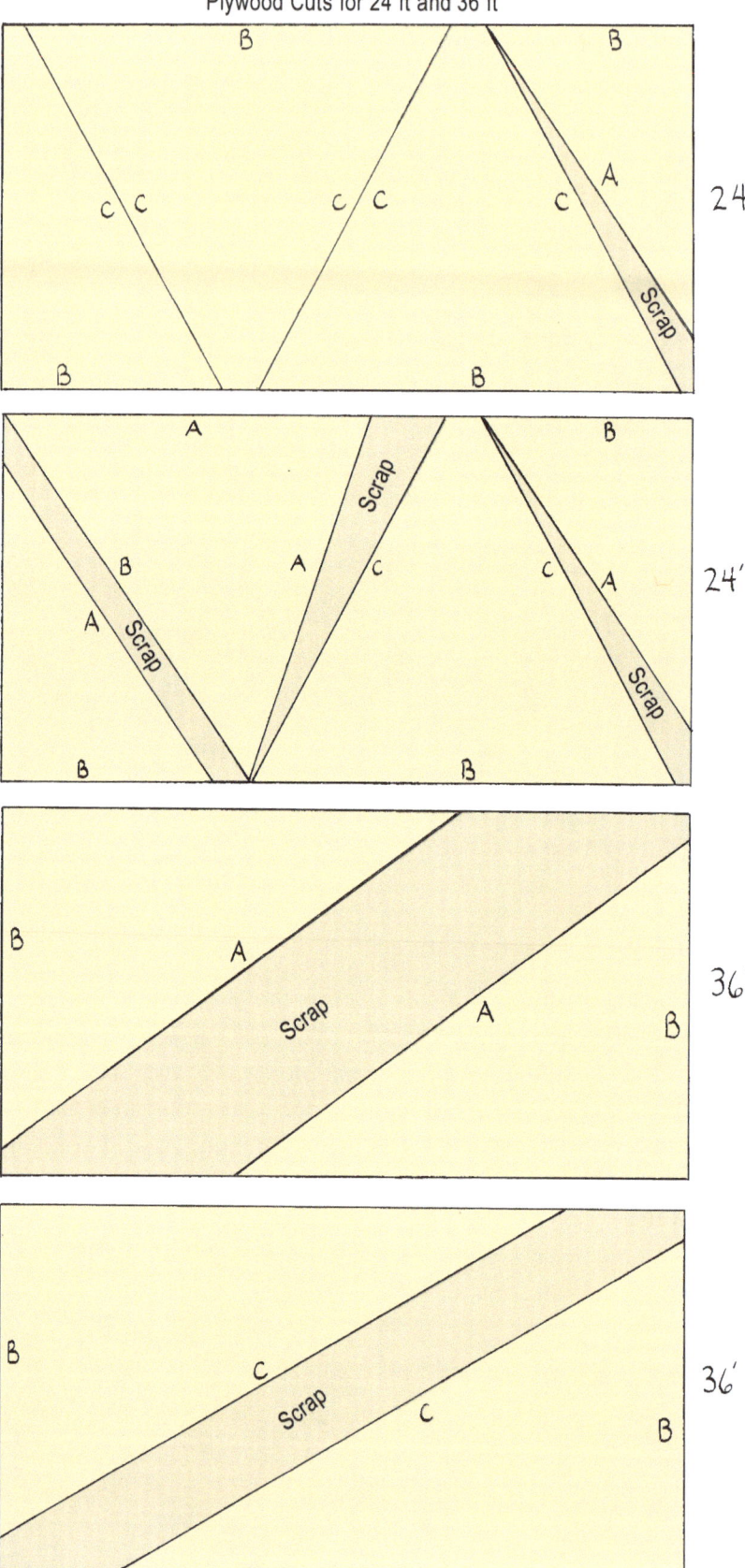

Plywood Cuts for 24 ft and 36 ft

The plywood triangles were cut several at a time by carefully stacking the sheets and temporarily nailing them together, marking the top one as a pattern and cutting through six layers at once using a worm-drive saw. Each dome had 105 triangles total, using two different shapes. There were 30 AAB triangles for the pentagons and 75 CCB triangles for the hexagons, so for two domes there were 210 triangles in all. Of those, 25 had windows or doors, so there were 185 plywood triangles cut for the sheathing, and hand-nailed to the frames. This was a massive job, which took several days, and created a new sense of urgency. The plywood needed to be covered immediately with plastic and tar-paper in order to protect it from rain, and possible de-lamination. Note: Some of the scraps of plywood were large enough to use in other places, while some were not.

Plywood sheathing

Stapling the plastic film on the domes was almost nightmarish, because the film was in large sheets, and the wind was gusty for days. Compounding the problem, we were using ropes hung from the top pentagon to scale the outside of the dome, because scaffolding and ladders were useless after the first level up due to the curvature of the shell. Using a hand stapler while hanging on to a rope and trying to keep the plastic down in the wind was a difficult and frustrating task. Somewhere early in this process I decided to invest in a compressor and a couple of air-driven staple-guns, which helped immensely, because the air hoses could be used to hold down the plastic, and the stapling did not require any squeezing action on the trigger. Simply holding down the trigger and bumping the head fired the staples repeatedly, so we were able to go much faster. Tar-paper was then stapled over the plastic vapor barrier, starting at the bottom and working our way up, overlapping as we went. Finally, several layers of 1-inch mesh chicken-wire were stapled on top of the tar-paper. The wire would give the gunite (masonry cement applied by spraying) something to stick to when it was wet, and after curing the wire became a reinforcing matrix. We had stapled over a half a million staples by this point!

Tar-paper over sheet plastic

Chickenwire over tar-paper

Framing the garage and entryway

Another cost-saving opportunity presented itself when the concrete subcontractor could not work it into the schedule to pour the front entry-way porch. We decided to do it ourselves, and after having observed the previous concrete pours including the back patio I felt confident that we could do the job. The forms were staked in place and reinforcing wire was unrolled inside them. I wanted an exposed aggregate look to the concrete, and this was done by ordering a specific mix from the plant that had small round pebbles in it. Once delivered and poured, we spread and leveled it to the top of the forms, but did not tamp it down. This left the pebbles in the top layer of concrete, which were then exposed by gently spraying with a water hose before it cured. The result was a nice pebbled look to the front porch.

Next we framed the garage and entry area, which took longer than assembling the 24-foot dome. The original plans called for a flat garage roof and overhang with a horizontal 6-foot wall in front of the courtyard. Again, this was not a practical design with the heavy snows that occur occasionally in Lubbock, not to mention the rain. It also lacked aesthetic appeal and so we made the adjustment to the plans. Every change to the plans required going back to the building inspection department and getting approval. The garage was designed to help transition the domes to the neighborhood in an aesthetically pleasing look as well as being more functional weather-wise.

Interior framing was time-consuming

    Having protected the plywood shell and having poured the other foundations and driveway, we turned our attention to finishing the framing of the interior walls, which was very time consuming due to the different angles. Wherever an interior wall met the interior of the dome's wall angles had to be measured, and every single one of them was different. There were no shortcuts that could be taken and this aspect of the project was tedious. We installed 2x6s into the window openings as part of the window trim treatment which would also give the framework additional support during the gunite process. Wet masonry cement would add <u>tons</u> of dead load to the framework until it cured, and I did not want to take any chance of a collapse. At that point I analyzed and reanalyzed the steps we had taken until I was sure it would not fail. The engineering professor had intended to load test the frame at that point of construction by hanging 55-gallon drums filled with water from the struts and measuring the amount of deflection from the slab to the top of the dome. By the time we finished framing the interior walls this was not possible, in fact not even necessary because the dome had become totally supported by the sub-frame, plywood sheathing, 2x6 window frames, and interior framing which extended to the very top of the dome, locking it all in place. (Even the three main struts cut away by an indifferent plumber installing three vent pipes did not weaken the structure, an incident that happened while I was not on site.) It was time to put the real load test on the structure, the masonry cement or stucco, which weighed 20 tons.

Scratch coat of stucco

A friend who was a swimming pool contractor was hired to stucco the shells and basement walls, using a machine that pumped the cement through a hose to the spray gun, hence the name gunite. Since it was impractical for him to scale the dome with a rope and spray cement, he set up scaffolding around the perimeter of the structure and bridged across them with long 2x12's, creating catwalks. This allowed him and his crew to spray the cement and smooth it as they went around the dome, starting at the top and working their way down. The gunite machine was perfect for applying the thick mortar mix because it filled the spaces between the chicken-wire very effectively, better than trying to hand trowel. As the spray gun operator would deposit the material, his helpers would smooth it over and score the surface as they went. Two layers were applied about an inch thick, called a scratch coat, followed by a final surface coat a couple of days later that had a sandy beige coloring agent in it. The final coat also had a texture to it, which helped to hide any imperfections. Two domes took about a week, during which time I kept the stucco moist by spraying it daily from a water hose to insure a proper cure. After several weeks, 55 gallons of silicone-based waterproofing agent were applied using a mop, and the shell was done! (I had started to use a water sealant sold in 1-gallon containers at the local hardware store, but as soon as I applied a little it began to noticeably stain and darken the stucco. It was on the front of the building on one triangle and was apparent for some time, but eventually faded, to my relief. When in doubt, try products in an inconspicuous place.)

Final coat of stucco with coloring agent

Not too long after this an enthusiastic realtor came by and wanted to sell the dome. I was told that there was a potential buyer, even though the project wasn't finished, so I decided to list the property. That was a mistake. In effect, this froze construction, because the new buyer would dictate how the work would be completed. When no buyer materialized the realtor suggested that we should have an open house some weekend, which I went along with hesitantly. I had reservations about how people would see and appreciate the dome in this unfinished state. The interior at that point had no insulation, sheetrock, interior doors or cabinets. An ad proclaiming 'Moon House for Sale' was run in the paper, much to my dismay and the open house was held on a Sunday. As people walked through the door the looks on their faces confirmed my suspicions. Without insulation or sheet-rock, and bare wood, plumbing and wiring showing, it was hard for most people to visualize the finished product. Needless to say, we did not sell the dome that day, or any other day during the listing. After the listing with the realtor expired, work resumed on the project.

Exterior doors and windows were the next step. There were only two exterior doors on the big dome, one at the entryway and the other going out of the kitchen onto the side patio. The small dome had one double door. The entryway door was a basic foam-filled metal door, and the kitchen had a glass slider. All of the doors were set in from the shell, each in an alcove design, and one of the problems was a lack of headroom going through one of the triangles into this area. (I didn't want to cut into the bottom support triangles.) There seemed to be sufficient room at first, but stucco and the concrete slab which was poured later seemed to significantly infringe on the size of the opening. Once 'written' in concrete it was too late to change. In hindsight, making the openings larger and reinforcing the surrounding struts with metal plates would have been the solution.

Front entryway

There were 15 windows on the big dome and 5 on the small one. The very top of both domes had pentagons which were skylights, and the problems to overcome included hail, rain, and the summer sun. The solution was 5/8-inch thick 'Solar Bronze' Plexiglas®, which was also used on the rest of the windows that were above ground level. Solar Bronze Plexiglas® would absorb some percentage of the heat from the sun instead of passing it into the house, but it was expensive. It also had a high expansion/contraction coefficient. The rest of the lower windows were regular 1/4-inch mirrored plate glass. All of the glass was custom cut, an additional expense. Since the windows had no frames to hold them into the triangular openings, we had to figure a way to make them work. The solution was fairly simple.

Window openings

A 2x6 triangular frame had been built for each window, which fit inside the 2x4 openings, and strengthened each section. (This step had been done before the stucco was applied, in order to run the masonry cement flush with the outside of the 2x6.) Inside this frame a 1x2 inch wide triangular frame was recessed to the thickness of the glass and finish-nailed into place. This was the backing that the glass would rest on, making the glass flush with the outside of the dome. We carefully lifted the plastic-masked Plexiglas® panels up the scaffold, angling them through the openings. Once they were in place we immediately peeled off the plastic masking which would have been difficult to remove once the heat baked it on. Massive amounts of silicone rubber caulk were used to seal the windows. The caulk was extended over the edge of the 2x6 frame and onto the stucco about a half inch, creating a band about 3 inches wide all around each window. Silicone caulk was a relatively new product at

Mirrored glass on bottom level - Note the silicone

5/8 inch thick skylights

that time, so product information was sketchy, but it seemed to do a good job as a sealant. (It does not have a 30-year life as claimed, however!) Again, the drawback was the dust sticking to it. In fact the dust storms in Lubbock carried a good deal of static electricity, and would transfer some of this charge to the Plexiglas®. As a result the windows attracted dirt like magnets, making them hard to keep clean. Plexiglas® scratches easily, so additional care had to be taken when cleaning them. Skylights were a nice design touch, but maintenance proved to be quite a hassle. Another unforeseen problem was that the expansion and contraction of the Plexiglas® (a half inch on a hot day) made a popping noise occasionally as they touched the wooden framework, and since sound carries well in a dome it was very noticeable. They should have been seated on a thin flexible material, but once installed there was no way to fix it short of taking the windows out. In fact, any sound at all seemed louder, and a heavy rain made a thunderous noise. You could hear a whisper if you were standing in the exact center of the dome. This problem improved tremendously after the insulation and sheetrock were done, which was the next step after the wiring was completed.

Scaffold work on the skylights

Basement staircase

One of the more aesthetically pleasing design features of the house was a 4-foot diameter spiral staircase that went from the basement to the upstairs loft. It was made from cast aluminum, and came in 4 boxes as a kit that required some assembly. It was well designed, sturdy, and bolted together with no problems. (In hindsight, a 6-foot diameter would have been better, as it would have made it easier to move furniture around, but it was significantly more expensive. Also the entry hole into the basement was only 4 foot in diameter.) Later, the staircase would be painted with a special epoxy paint that would stick top aluminum and the rubber railing cover would be installed. A custom metal railing was fabricated at a welding shop for the top landing.

The spiral staircase required a special epoxy paint, because it was made from aluminum and regular paints don't adhere well to aluminum. Also, durability was important. Here's a warning: It takes a respirator with a special filter cartridge to handle epoxy mists! Xylene was the solvent used to clean out the spray gun, also hazardous!

Railing on top landing

Staircase and small dining area next to windows overlooking the patio

View of fireplace and loft overhang on the right

Another nice design touch was a free-standing metal fireplace (also a kit) sitting on a hearth of white marble, and was capable of heating the entire dome. It proved invaluable working inside during the winter, before the heating/cooling unit was installed. The wall behind the fireplace was stuccoed all the way up to the top of the chimney, reflecting the heat back inside. The fireplace design created a draft of air which made a tornado of flame, visible through tempered glass. The fireplace was not originally in the plans, so a hole had to be cut in the dome to accommodate the chimney pipe. Using a plumb bob we established the exact location, marked a circle on the plywood sheathing inside the dome, and drilled a series of holes around that circle using a masonry bit. The stucco was about an inch and a half thick in that spot, so it took a little while. After this was done a small sledge hammer was used to break the circular chunk out, and wire cutters took care of the chicken-wire. Sections of single wall chimney pipe were bolted together, and extended through the roof to a height of about 10 feet, for good draw. Silicone rubber caulk was used to fill the gap between the dome and pipe, because it could withstand the heat. It all worked very well except for one problem. Single wall chimney pipe condensed moisture, both inside the dome and outside, eventually rusting the pipe and staining the stucco as it dripped. It did not seem like a major problem at the time, but over several years this process rusted out the fireplace. Another small problem with the fireplace: the dome was so tightly sealed that when the fireplace was being used, caution had to be taken when closing the front door suddenly. If the front door closed quickly, one of the fireplace glass panels would break, imploding into the fireplace. It was a special tempered glass, and a nuisance to replace.

Stained concrete on back of dome

Fireplace on white marble hearth

Angled roof flows into privacy wall

At that point of construction the exterior had been finished. An angled brick privacy wall had been built across the front of the house. It tapered down from the garage to the smaller dome, creating an entry courtyard. This was done for two reasons. The first was to help the radical structures blend in with the existing houses a little better, and the second was to try to control the constant flow of tourists wandering through the property. On the day that the brick wall was finished I thought we had the problem under control, but when I returned later that evening I noticed that several bricks were down off the uncured wall! Upon going into the house, I saw a couple of people in the back yard wandering around. They had climbed the wall, still wet uncured mortar, damaging it as they went over. It was at that point I realized there would be an ongoing problem with tourists even after the house was finished. (Subsequent owners have reported instances where people would just walk in the house unannounced!) If you build a dome, they will come...

Entry

Six-inch thick fiberglass insulation was used instead of four inch to help with the noise problem, and was stapled to the plywood sheathing. The exterior walls only had 3½ inches of cavity space, but the fiberglass would compress that much. Supposedly it was not a good thing to do that, but I found this added to the efficiency in heating and cooling the structure, which was verified by the first utility bill after AC installation. One small problem associated with thicker insulation was that it was harder to sheet-rock over it, since it stuck out more. This problem was compounded by the weight of the sheet-rock (it is heavy) and the fact that most of it had to be put up using scaffolding. Each triangle had to be cut individually since you can't mass-cut it, which was very time-consuming and then hand-carried up the scaffolding. It used the same basic pattern as the exterior sheathing and so we had a fair amount of wasted sheetrock.

Another small problem emerged, which was very aggravating. Some of the sheet-rock nails that I bought were not galvanized, and I did not realize it at the time. When the taping and floating of the joints was done, small rust spots bled through everywhere that there was an un-galvanized nail, necessitating sealing each and every one with varnish to prevent them from bleeding rust spots through when the painting was done. After all this, acoustic ceiling texture was sprayed on, starting above the first level of triangles at ground level. This acoustic material helped abate the noise problem a lot. The rest of the house had a simple spatter texture on all walls accessible to traffic, except for the entry hallway, where brick facia was used for extra durability. The brick facia was 3/8-inch thick and was painstakingly glued to the walls. A few other interior walls had pecan paneling over the sheetrock, and the upstairs bathroom had shiplap boards on the exterior of the bathroom.

Kitchen and bathroom cabinets were the next phase of construction, and were built by a friend in his garage. Ash was chosen for the shelves, drawers, and door fronts because of its strength and nice grain. Shelves made of 3/4-inch ash plywood would resist sagging better than most other materials, and was the cabinet maker's preferred wood, so I left the design of the cabinets up to him. The oven, cooktop and vent hood were chosen at that time since he needed the dimensions, and were state of the art, with electronic touch pads and a solid ceramic top. Since the kitchen was small it needed to be efficient, and his design took advantage of every square inch. Once assembled, I stained and varnished them a couple of times, sanding between coats with fine steel wool. Next, the white Formica counter top was carefully glued on with contact cement and trimmed with a router and finally, the door and drawer handles were installed. The same process was repeated for the bathroom cabinets, with the same care and attention to detail.

Kitchen cabinets, cooktop and oven

Downstairs bathroom

Kitchen cooking area and doorway to utility room

The mechanical room where the air conditioning and heating air handler was located in a small utility room behind the kitchen cooking area. Duct work for the air conditioner ran through the bottom of the upstairs floor, which was constructed of 2x12s. This allowed adequate room to run the ducts, wiring and plumbing, but it was a challenge.

Wiring was done by a licensed electrician, and did not take long. One problem that we had (other than extra expense) was an inability to run wiring to a floor plug due to a blockage in the conduit, which happened during the pouring of the slab. Floor plugs are nice, but they involve more work and additional cost, and in this case it was for nothing.

In fact, I began to realize that at every single phase of construction (except the stucco, which was a great deal) the dome was costing more from the subcontractors than a regular house would! There should have been no difference.

There were other minor problems, like the location of the downstairs toilet drain being too close to the wall. (Remember the importance of checking the subcontractor's measurements?) The solution was that a special toilet had to be ordered, an additional expense. Problems with the plumbers seemed to be a constant, like when they told me that the sewer line in the alley was higher than they had figured, so I would need an expensive lifting pump, around $4000. I opted to not go that route, looking for an alternative solution. Soon after that they went out of business, which was lucky for me, because the new plumbers that I hired worked relatively problem free. The solution to the lifting pump was simply to raise the drain line to coincide with the sewer line in the alley. It worked, and with virtually no expense.

Kitchen sink and dishwasher

Installation of the sinks, toilets, upstairs shower stall and downstairs tub was the next series of jobs, and I was able to do them without the aid or expense of a plumber. I did invest in a series of 'How To'…books, which were invaluable. Sinks were installed by cutting holes in the countertops, setting the sinks in place on caulk, bolting on the faucets, and hooking them up to the existing supply lines with compression fittings, and attaching the plastic drain and trap. Toilets were assembled, tanks to bowls, and set in place over the drain with a wax ring in between, bolted to the drain flange, and hooked to existing supply lines with compression fittings. Hooking up the tub and shower was a little trickier, because pipes had to be soldered together using a torch, which took a little practice. I simply went by the book. It was occasionally very frustrating, though, because doing something for the first time often means you have to do it twice to get it right. Each phase of construction was like completing a new course of study in school by this point, and graduation would be the completion of the entire building project. Once the plumbing fixtures class was done it was time to enroll in another class, like the "tiling the tub and shower stall" class.

Custom cabinet in master bath

After reading the book, I skipped that class and hired a tile installer to do the shower and tub because it was relatively inexpensive. After watching him I was glad I did, because there were additional angles, not just flat walls and 90˚ angles on the downstairs shower-tub enclosure. It required precise angle cuts with a tile cutting saw to make it look right. Once the upstairs shower was tiled, I installed the glass door unit, caulked, and it was finished. A nice design touch was that one of the skylight triangles in the pentagon at the top of the dome was over the shower stall.

Tile angles on dome

By the time we had gotten to this phase of construction

Skylight over the master bath shower stall

more than a couple of years had passed, due to the 6 months time off when the house was listed with the realtor, and a general loss of enthusiasm for the project. Even though the house was nearing completion it seemed as though it would never be finished. Still, one by one the final details were done, like the suspended acoustic ceiling in the basement.

Acoustic ceiling in basement

An interlocking grid of metal framework was hung from wires which were hooked on nails hammered into the bottom side of the wooden forms in the basement ceiling which were left in place after the concrete slab was poured. It was easy to level up the framework by measuring off the floor and adjusting the wires, and the final step was to just put the lightweight ceiling tiles into the grid, along with recessed lighting fixtures. The entire ceiling only took about 6 hours, was relatively inexpensive and looked good. It also helped tremendously with the sound echoing off the hard cinder-block walls and floor.

The interior doors, paneling and other trim work came next, taking several weeks due to the variety of angles we encountered. Paneling was used in the downstairs bedroom and any other walls that hadn't been textured, like the walls under the loft, and was glued with Liquid Nails® to the sheet-rock. It was also nailed with an air-driven finish nail gun, as was the trim. Wooden doors that had been hung were sprayed with varnish, and the knobs were installed. Finally, it was time for the final step, painting. I bought a 2-gallon paint pot that was air driven, long hoses and a spray-gun, several 5-gallon buckets of acrylic latex, and proceeded to spray the interior. Note: Clean the paint spraying equipment thoroughly after each use, and it will last for years.

Hallway with faux brick and paneling

Painting around the Plexiglas® windows presented a problem. Because it scratches easily, the use

Entry hallway with faux brick

Faux brick was glued to the sheetrock in the entry hallway and near the kitchen sliding door, providing a wear-resistant finish in high-traffic areas. This proved to be very tedious and time consuming, taking about 10 hours.

View from front door into the living room

Bedroom hallway with faux brick

Skylight

of a razor blade to remove any over-spray was out of the question, so they had to be fully masked off so as to not get a single drop of paint on them. Compounding the problem was the fact that after only a couple of days the masking tape did not come off cleanly, because the heat from the sun cooked the tape onto the glass. Methyl alcohol and a lot of 'elbow grease' had to be used to remove the tape residue, since other solvents would have damaged the glass.

After the painting was finished the dome was ready to have the flooring installed. Carpet was

Shiplap siding creates visual interest, with connecting plant ledge

Railing on second floor

chosen for the basement, main living area and upstairs loft, with linoleum going in the entry, kitchen and downstairs bathroom. Tack-strip was nailed around the walls to hold the carpet in place, and foam padding was glued to the floor. Finally, the carpet was installed, starting with the basement. Indoor-outdoor carpet with a foam backing was used in the basement and glued directly to the floor, so no tack strip was necessary. The noise level was noticeably reduced, but there was still some echo effect from the hard block walls. Once all the carpet was laid, the noise level in the whole house went down by at least 10 decibels. It also reduced the noise going up and down the metal staircase. The only problem with carpeting was again the waste factor and expense because of the curved walls, but still, it really was the finishing touch. All that remained was some repainting on the baseboards, which had taken some scratches and dings during this final step, and for the most part the dome was finished.

The final inspection was the last step in the building

Carpeted stairs

Basement staircase

Fireplace with stucco wall behind

Silicone sealant on windows

process, and I suspected that there would be a problem. Sure enough, the building inspection department told me they could not give the final approval because some of the scheduled inspections like electrical and plumbing had not been done, since they showed no entry into their logbook! Fortunately though, I had saved all of the green tags from those inspections, and when I produced them they had to comply. Note to anyone wishing to build a dome or any structure inside the city limits of any town: SAVE YOUR GREEN TAGS, because apparently the inspections don't always get logged! Building inspectors get busy sometimes, and forget to record it back at the office.

Once finished, there were plenty of eager buyers for the house, because it was aesthetically pleasing, unique, and well built. Actually, it was overbuilt. The problem was that all the buyers needed to finance the house through banks or other lending institutions, and every single one of them refused to loan money on it! Their rational was that it was an unproven structure, and might not last 30 years. So, if you've ever wondered why there aren't more progressive types of structures being built, and why the forests are disappearing, there is a reason. Fortunately, I had one serious buyer who managed to secure private financing for the dome, and the sale went through. After that, plans to build more domes were put on the shelf, and the experience became one of artistic statement rather than ongoing commercial enterprise.

The aesthetics of the structure had determined every aspect of construction, as though it had a mind of its own, and the end product bore little resemblance to the original plan. It was, in fact, a large piece of sculpture that had been modified and refined in much the same manner as any artwork, to its ultimate final form.

The floor plan of the house was done in such a way so as to leave as much open space as possible. As you entered the front door there was a short hallway leading directly into the main living room, which was open to the top of the dome and the skylights. To the left, around the spiral staircase, was the kitchen. To the right was the fireplace

Silicone sealant on windows

and a short hallway leading into the downstairs bedroom and bath. The loft was separated from the living room by a half wall, and the only enclosed room upstairs was the bathroom and closet. The loft bathroom was pie shaped, the apex being at the center of the dome, which gave it the shape of the bow of a ship. Inside that bow was the shower stall, with skylight. Horizontal tongue and groove paneling enhanced the 'ship' effect.

Patio door from kitchen

In retrospect, too much attention was given to aesthetics and not enough to functionality, so as a result there was a lot of wasted space in some areas, and not enough space in others. The openness of the interior was nice, but gave no privacy to the upstairs area. A more functional use of the space would have been to completely separate the downstairs from the upstairs, which would have given more square footage of floor space, increased privacy upstairs, and would have eliminated a lot of noise and echo effect.

Other improvements in the design and construction of the dome have been realized since its completion, some from personal observations and the rest from feedback passed on by current residents. One element that was a nice design feature but not very practical in the long run was the skylights, for several reasons. First, as previously mentioned, they popped with changes in temperature. Second, they attracted dirt. Third, they were hard to clean. Fourth, they condensed moisture inside the house on cold days. Fifth, they allowed too much heat in during the summer. Sixth, and this was a major headache, they eventually leaked! The problem was that the silicone rubber dried out and lost elasticity, and the constant expansion and contraction of the Plexiglas® finally pulled it away from the caulk. I've seen this happen within four years, even using premium 50-year silicone caulk. The problem with the caulk at that point was that it had not even been around for 50 years and so product life was mere speculation. (I used premium 30-year silicone caulk several years later on a greenroom project, only to have it fail in 3 years. The manufacturer required proof of usage in the form of returned empty cartridges! Who saves the empty cartridges?) The reason for having skylights is outweighed by the negative factors in the long run. The only way to maintain the waterproof integrity of the dome is to eliminate unprotected skylights, or to be prepared to caulk every year. Another alternative, if skylights are a must, is to have them prefabricated with flanges and flashing that is embedded in the stucco.

Patio doorway into kitchen

In fact, the standard concrete shell itself needs attention every couple of years or so, i.e. a waterproofing agent, because of constant exposure to the elements and the tendency for stucco to develop hairline cracks. There is, however, a better method of construction now available that would probably eliminate these problems, which involves the use of polymers and fibers added to the masonry cement during the original guniting process. In fact, even the multiple layers of chicken-wire could probably be reduced to a single layer, since the fibers in the mix would take their place.

Polymerized concrete has been proven to be stronger, more waterproof and to last longer than regular concrete and is now being used in structural applications. If I were to build a dome today, this would definitely be the way to go.

Some other feedback that I've received concerned the lack of headroom around the doors. At that time I was reluctant to take out too many of the bottom struts in order to have a larger entrance way, because I felt that it might compromise the strength of the frame during the guniting phase. However, since both doors were set back inside alcoves and the interior framing and upstairs floor joists were attached to the dome frame around the door areas, there was plenty of reinforcement in place which would have allowed the removal of several struts. Hindsight is 20/20.

So now you've finished this book and are thinking about building a geodesic dome.
Here are some suggestions.
1. Read several books on domes and also books on conventional construction.
2. Draw up your plans, and then try to visualize walking through the structure.
3. Build small models to further familiarize yourself with the structure.
4. Work out a realistic budget, and assume that everything will cost more. Small domes cost less.
5. Limit extravagances, and once you have started building avoid too many changes, especially with subcontractors.
6. Talk to your banker or finance company first. They may or may not be willing to work with you on financing your project. If they won't, you'll need private funding.
7. Have blueprints drawn up by a reputable architect, especially if you are building inside the city limits.
8. Talk to the building inspection department first if you plan on building inside the city limits.
9. Build outside the city limits if at all possible.
10. Do as much of the work yourself as you can.
11. Obtain several bids from different subcontractors, and spell out exactly what is expected.
12. Try to remain flexible and to look at the project as an exercise in problem solving.
13. Research everything.

I would urge the reader who is contemplating building a geodesic dome to search out other information, especially people who have actually had a first-hand experience with the structure and to use this knowledge as a base to build upon. With a little planning and persistence, and a vision of what you would like for it to be, it can be done.

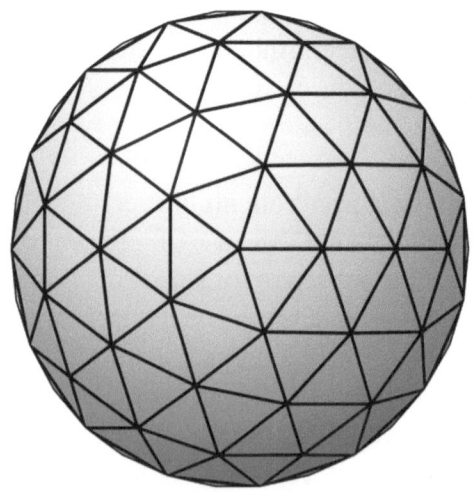

Master bath cabinets

View from the street

24-foot dome interior

Front porch entry

24-foot greenhouse

Fireplace from landing

Native plants were used to help the domes blend in

Plexiglas® skylight faced southeast

www.ingramcontent.com/pod-product-compliance
Lightning Source LLC
Chambersburg PA
CBHW051105180526
45172CB00002B/786